Brock
BIOLOGY *of*
MICROORGANISMS

Eighth Edition

Study Guide
Robert E. Andrews
Iowa State University

Madigan   Martinko   Parker

Acquisition Editor: *David Brake*
Production Editor: *Kim Dellas*
Special Projects Manager: *Barbara A. Murray*
Production Coordinator/ Buyer: *Ben Smith*
Supplement Cover Manager: *Paul Gourhan*

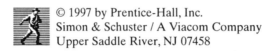

© 1997 by Prentice-Hall, Inc.
Simon & Schuster / A Viacom Company
Upper Saddle River, NJ 07458

Printed in the United States of America

10 9 8 7 6 5 4

ISBN 0-13-533712-7

Prentice-Hall International (UK) Limited, *London*
Prentice-Hall of Australia Pty. Limited, *Sydney*
Prentice-Hall Canada, Inc., *Toronto*
Prentice-Hall Hispanoamericana, S.A., *Mexico*
Prentice-Hall of India Private Limited, *New Delhi*
Prentice-Hall of Japan, Inc., *Tokyo*
Simon & Schuster Asia Pte. Ltd., *Singapore*
Editora Prentice-Hall do Brasil, Ltda., *Rio de Janeiro*

TABLE OF CONTENTS

PREFACE

This study guide is a companion to the text *Brock Biology of Microorganisms*, eighth edition by Madigan, Martinko, and Parker. The original hard copy form was written by Allan Konopka (Purdue University), and was revised for the seventh edition by Ronald Turco (Purdue University). The study guide was revised for the eighth edition and prepared for on-line presentation by Robert Andrews (Iowa State University). Chapters in the study guide have been numbered to correspond with those in the main text.

The study guide contains an **Introduction** including a chapter overview and chapter notes. The Introduction highlights most of the important concepts in the chapter. **The Key Words and Phrases** section reviews the major vocabulary to be gained in the chapter and includes text references. The **Multiple Choice** section contains a self-test that will be graded for the student. The **Discussion** section contains a series of essay questions. **Answers** for all questions are found at the end of each chapter.

Microbiology has been an important science for the past 100 years in that it has provided the means to control a number of infectious diseases and the experimental systems for the development of molecular biology. New developments in biotechnology and environmental microbiology indicate that microbiology will continue to be an exciting field of study in the future, and it is our hope that this study guide is an aid to your learning about this science.

Chapter 1
Introduction: An Overview of Microbiology and Cell Biology

OVERVIEW

Chapter 1 (pages 1-32) provides descriptions of the biology of cells and the history of microbiology. Even though the science of microbiology has developed only in the last 100 years, it has been a very important science for two reasons: (1) microorganisms have been excellent research tools for understanding the molecular biology of cells, and (2) many problems important to human society are consequences of microbial activity.

CHAPTER NOTES

All living organisms are composed of **cells**. Five important characteristics of all cells are self-feeding or nutrition (the "machine" function), self-replication or growth (the "coding" function), differentiation (form new cell structures such as spores as part of their life style), chemical signaling (communication with other cells) and evolution (change to show new biological properties). Superficially, cells seem to disobey a law of physics -- they are highly ordered structures in a world that generally becomes less ordered with time. How do they maintain order? By continuously generating energy, some of which is used to maintain cell structure. Energy generation is one important component of **metabolism**; other aspects of metabolism include the chemical reactions that synthesize the compounds and assembly reactions that make up the cell structure. Chemical reactions in cells are catalyzed by protein molecules called **enzymes**. Enzymes must have a specific structure to function; therefore, there must be a set of information (a gene) that encodes the structure of each protein in the cell. The instruction set is encoded in **DNA**, the genetic material of all cells. There is also a translation system **RNA**, to convert the information coded in the DNA to proteins. Several types of RNA molecules (messenger RNA, ribosomal RNA, and transfer RNA) are important in this process.

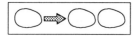

1. Self-feeding (nutrition)
Uptake of chemicals from the environment and elimina of wastes into the environment.

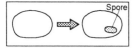

2. Self-replication (growth)
Chemicals from the environment are turned into new ce under the direction of preexisting cells.

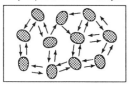

3. Differentiation
Formation of a new cell structure such as a spore, usually as part of a cellular life cycle

4. Chemical signaling
Cells communicate or interact promarily by means of chemicals which are released or taken up.

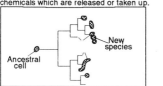

5. Evolution
Cells evolve to display new biological properties. Phylogenetic trees show the evolutionary relationships between cells.

The result of biosynthesis is cell **growth**. For a cell to replicate itself, it must synthesize more than 1000 different protein molecules. The cell has the genetic information to produce about 3000 distinct proteins; the genes that are expressed are those which encode proteins that are most useful for growth or survival under the existing environmental conditions.

The cell must also faithfully copy its genetic information, to pass onto the new cell. Mistakes in copying are made occasionally; these **mutations** are usually harmful and kill the cell. However, they do provide a mechanism for cells to acquire new properties. This occurs if the protein coded by the mutated gene catalyzes a different reaction than the original protein. Under the appropriate environmental conditions, this mutated cell can have a selective advantage (that is, it can replicate faster than its competitors). This principle of **natural selection** is the mechanism underlying Darwin's theory of **evolution**.

There are two structural types of cells: prokaryotic cells are relatively simple in structure; eukaryotic cells are more complex, in that they contain **organelles** that are compartments for special metabolic functions. These organelles include a true nucleus, the mitochondrion, and the chloroplast. In addition, biologists also deal with **viruses**, which are non-cellular entities that use the metabolic machinery of cells to replicate themselves. The dichotomy in the structural types of cells does not accurately represent the evolutionary relationships among organisms. Analysis of the nucleotide sequences of ribosomal RNA has indicated that there are two groups of prokaryotes: the **Archaea** and the **Bacteria**. These groups are no more closely related to each other than they are to the **Eukarya** (see figure below).

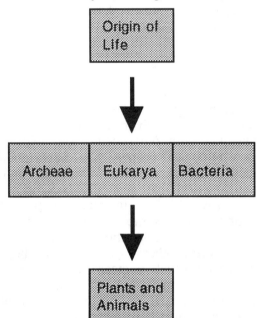

Microorganisms live in natural habitats in which their growth is affected by interactions with **populations** of other microbes, as well as by the physical and chemical characteristics of their environment. Understanding the ecological interactions in microbial communities is extremely important to determine the role of microbes in nature. It is difficult to study microbes in nature; therefore, most of what we know about microbes has been learned from **pure cultures**.

Some microbial species can have devastating effects on human beings by causing **infectious diseases**. A great success of the science of microbiology has been the control of fatal infectious diseases in developed countries. However, these diseases are still important causes of death in less developed parts of the world. Despite these threats by some species, most microorganisms are beneficial. The proper function of the **biosphere** and soil depends upon their activities. Of more direct impact upon humans is the industrial production of antibiotics, food products, organic chemicals, and biomass. A recent development has been the biotechnology industry, which uses microbes as factories to produce proteins from animal and plant genes that were introduced into bacterial DNA. Microbes are also important in agriculture and food spoilage.

Microbes were discovered 300 years ago by the microscopist **van Leeuwenhoek**, but most developments in microbiology have occurred in the past 100 years. **Louis Pasteur** and **Robert Koch** were two leaders in developing the discipline. Among many other accomplishments, Pasteur disproved the theory of **spontaneous generation** conclusively. This was important to demonstrate that experiments with microbes were reproducible, and not the consequence of new life arising. Koch developed rigid criteria for proving a specific disease was caused by a specific bacterium (**Koch's postulates**). Obtaining pure cultures of microbes was essential for fulfilling Koch's criteria and the use of pure cultures remains an essential tool in the study of infectious disease today.

SELF TESTS

COMPLETION

1. Bacteria are named using a binomial system. The first letter of the _____ name is capitalized, the first letter of the _____ name is spelled with a lower case letter, and both words are _____.

2. The 3 groups of eukaryotic microorganisms are_____, _____, and _____.

3. Bacteria, Archaea and Eukarya can be differentiated by comparing _____.

4. _____ are non-cellular entities that are studied by microbiologists.

5. All living organisms have as basic structural units _____.

6. An organism in which the genetic material is not bounded by a double membrane is a _____ organism and its nuclear region is called the _____.

7. Bacteria range in size from _____.

8. Enzymes are _____ molecules that _____ chemical reactions.

9. Before cells divide, they must _____ their genetic material.

10. Microbial cells live in assemblages called _____ and associate with other cells forming a _____.

11. To synthesize a specific protein, the coding information in DNA must be transcribed by synthesizing a molecule of _____. This molecule is then _____ into a sequence of amino acids by _____.

12. Food products that require microbial metabolism for their manufacture include _____, _____, _____, _____, and _____.

13. _____ was the first to show the role of microorganisms in human diseases.

14. The complex mixture within the cell is known as _____ and it is held in place by the _____ which is surrounded by a _____.

15. One *E. coli* cell contains about _____ base pairs, _____ protein molecules, and _____ kinds of protein.

KEY WORDS AND PHRASES

Antoni van Leeuwenhoek (p 20)	Archaea (p 12)
aseptic technique (p 16)	Bacteria (p 12)
binomial nomenclature (p13)	biomass (p 20)
biotechnology (p 20)	cell (p 3)
cell growth (p 4)	cell membrane (p 3)
chemical signaling (p 4)	community (p 14)
culture medium (p 16)	cytoplasm (p 3)
differentiation (p 4)	DNA (p 7)
ecology (p 14)	ecology (p 14)
ecosystem (p13)	endospores (22)
enzymes (p 6)	Eukarya (p 12)
eukaryote (p 10)	evolution (p 9)
gene (p 8)	gene expression (p 10)

3

genetic code (p 7)	genetic engineering (p 20)
genetics (p 9)	genus (p 13)
habitat (p 20)	Koch's postulates (p 24)
Louis Pasteur (p 22)	metabolism (p 6)
microbiology (p 2)	mutation (p 9)
natural selection (p 9)	nucleoid (p 3)
nucleus (p 3; p10)	nutrition (p 4)
population (p 14)	prokaryote (p 10)
Protozoa (p 14)	pure culture (p 15)
RNA (p 7)	Robert Hooke (p 20)
Robert Koch (p 24)	self-replication (p 4)
species (p 13)	sterilization (p 16)
transcription (p 7)	transfer RNA (p 8)
translation (p 7)	virus (p 10)

MULTIPLE CHOICE

1. Prokaryotic cells contain all of the following except:

   (A) cell membrane   (D) cell wall

   (B) ribosomes   (E) DNA

   (C) nucleus

2. The process of translation involves:

   (A) messenger RNA   (D) more than one of the above

   (B) ribosomal RNA   (E) all of the above

   (C) transfer RNA

3. The diversity of life present on earth is a consequence of:

   (A) metabolism   (D) transcription

   (B) natural selection   (E) translation

   (C) replication

4. Which of the following groups is closely related to Bacteria?

   (A) algae   (D) fungi

   (B) Archaea   (E) none of the above

   (C) Eukarya

5. Microorganisms can have substantial effects upon natural environments because:

    (A) they are small   (D) they can differentiate

    (B) they can grow rapidly   (E) they have cell walls

    (C) they contain viruses

6. At the beginning of the twentieth century, the major cause of death in the United States was:

    (A) accidents   (D) infectious disease

    (B) cancer   (E) suicide

    (C) heart disease

7. Viruses:

    (A) cannot infect bacteria   (D) all of the above

    (B) are very small cells   (E) none of the above

    (C) have metabolism of their own

8. Microbial diversity is reflected in variations in:

    (A) cell size and morphology   (D) metabolic strategies

    (B) adaptation to environments   (E) all of the above

    (C) cell division

9. A simple Bacteria like *E. coli* will contain:

    (A) about 1900 kinds of proteins   (D) no proteins, it is prokaryotic

    (B) less than 400 kinds of proteins   (E) none of the above

    (C) a few proteins and a lot of enzymes

10. Cell walls are _____ than the cell membrane.

    (A) less flexible   (D) stronger

    (B) more porous   (E) all of the above

    (C) thicker

11. The only prokaryotes are:

    (A) fungi and virus   (D) virus and bacteria

    (B) Bacteria and Archaea   (E) Archaea and fungi

    (C) algae and animalcules

12. Bacterial functioning in soil and agricultural systems control what processes:

    (A) soil nutrient cycling   (D) formation of plant root nodules

    (B) plant nitrogen fixation   (E) all of the above

    (C) rumen digestion

13. Habitat describes what key feature about a population?

   (A) where a population lives

   (B) size of the population

   (C) the relatedness within a population

   (D) the cooperative nature of the population

   (E) all of the above

14. What were Bacteria and Archaea once known as:

   (A) bacterium

   (B) blue greens and cyanobacteria

   (C) Eukarya and Prokarya

   (D) eubacteria and archaebacteria

   (E) none of the above

15. What sets cells apart from non-living systems?

   1. self feeding

   2. growth

   3. differentiation

   4. chemical signaling

   5. evolution

   (A) 1,2,3; (B) 1,2,3,4,5; (C) 1,2,5; (D) 3,4,5 (E) 1,2,3,5.

FILL IN (Label the structures of the microbial cell.)

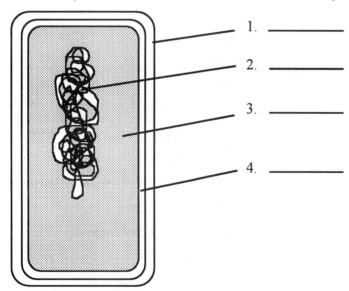

1. _____

2. _____

3. _____

4. _____

DISCUSSION

1. List three subdisciplines of microbiology that are oriented toward (a) understanding the role of microbes in natural environments or (b) solving an applied problem.

2. Name a major contribution to microbiology made by the following individuals: (a) Antoni van Leeuwenhoek, (b) Louis Pasteur, and (c) Robert Koch.

6

3. Scientists know that the disease AIDS is caused by a virus. What experimental procedures would satisfy Koch's postulates to prove that the virus was the causative agent of AIDS?

4. Discuss five characteristics by which living cells can be distinguished from strictly chemical systems.

5. Define the term metabolism.

6. Explain the why the small size of microorganisms does not render them helpless.

7. What was the theory of spontaneous generation? What experiment finally disproved the theory?

8. Bacterial strains have been found which are not killed by antibiotics although these chemicals had previously been effective in controlling the species. By what mechanism could such strains have arisen?

9. Explain why the study of microbiology has provided major advances in (a) basic science and (b) applied science.

10. Explain the significance of the principle enunciated by Virchow, "Every cell from a cell" in terms of the views of a cell as a chemical machine and as a coding machine.

ANSWERS

Completion

1. genus, species, italicized or underlined

2. algae, fungi, protozoa

3. ribosomal RNA sequences

4. viruses

5. cells

6. prokaryotic, nucleoid

7. 1 to 5 μm

8. protein, catalyze

9. replicate

10. populations, communities

11. messenger RNA, translated, ribosomes

12. cheese, yogurt, pickles, sauerkraut, alcoholic beverages

13. R. Hooke

14. cytoplasm, cytoplasmic membrane, cell wall

15. 4.6 million, 100,000, 1,900.

Multiple choice

1. C; 2. E; 3. B; 4. E; 5. B; 6. D; 7. E; 8. E; 9. A; 10. E; 11. B; 12. E; 13. A; 14. D; 15. B.

Fill in

1. cell wall; 2. nucleoid; 3. cytoplasm; 4. cytoplasmic membrane

Discussion

1. See text sections 1.6 and 1.8

2. See text section 1.9

3. See text section2 1.8 and 1.9

4. See text section 1.1

5. See text section 1.2

6. See text section 1.6

7. See text section 1.9

8. See text section 1.2

9. See Chapter Introduction and text section 1.9

10. See text section 1.1

CELL CHEMISTRY

OVERVIEW

Chapter 2 (pages 33-54) concerns itself with the chemical processes that allow cells to function. The chapter starts with the atom, goes through the processes that lead to the macromolecules that make up the cell structure, and ends with the concept of stereoisomersim.

CHAPTER NOTES

All forms of life contain similar types of **molecules**, and many of these biochemical compounds are not found in non-biological materials. Although they are unique, cellular molecules obey the laws of physics and chemistry. The molecules are composed of **atoms**; of the 92 natural elements on earth, only six are major components of biomass (Hydrogen, Carbon, Nitrogen, Oxygen, Phosphorus, Sulfur). **Carbon** is the most important of these; it can chemically bond to other atoms in a number of ways to produce diverse and complex molecules. Atoms consist of a nucleus that contains positively charged protons and neutrons, and a number of negatively charged **electrons** that orbit the nucleus.

1. Hydrogen bonding between water molecules

2. Hydrogen bonds between amino acids in protein chain

Cytosine   Guanine

Thymine   Adenine

Hydrogen bonds

3. Hydrogen bonds between bases in DNA

The sharing of electrons between atoms constitutes a **chemical bond**. If electrons are shared equally, the bond is **covalent** and is a relatively strong one. Covalent bonds can only be formed or broken by the specific reactions catalyzed by **enzymes**. Several non-covalent interactions are also biologically important in determining the shape of macromolecules, or the binding of macromolecules to other compounds. (1) **Hydrogen bonds** are formed by the spontaneous attraction of slightly positive to slightly negative atoms in molecules. A single H bond is much weaker than a covalent bond, but when a large number of H bonds are formed in or between macromolecules, they can have a large effect. (2) In **hydrophobic** interactions, molecules that are repelled by water associate with one another.

The nature of life on earth is determined by not only the chemistry of the carbon atom, but also the solvent properties of water. The greatest proportion of a cell's weight (70-90%) is due to its water content, and biochemical reactions will not occur unless there is an adequate amount of water available. Water molecules are slightly **polar** (that is, the + and - charges tend to be separated in the molecules). The consequences of this polarity are that (1) biochemically important polar molecules (such as proteins, nucleic acids, and the small molecules used as nutrients and building blocks) are soluble in water, and (2) nonpolar molecules (such as lipids) are not soluble in water and aggregate or

9

clump together. The aggregation of lipids form barriers such as membranes, and affects the movement of polar molecules into or out of cells.

Macromolecules form the cell's structure, catalyze its metabolic reactions, and contain the coding information to replicate a cell. There are four types of important cellular macromolecules: nucleic acids, proteins, polysaccharides and lipids. Each macromolecule is a **polymer** composed of a sequence of **building blocks** connected by covalent bonds. It is important to know the characteristic building block for each type of macromolecule. In nucleic acids and proteins, the **sequence** of building blocks varies among molecules; it is this sequence that determines the biological activity of these macromolecules in cells.

Polysaccharides are composed of **carbohydrate** chains ranging from 100 to several thousand units long. Carbohydrates containing 4-7 carbon atoms are most common in cells. Different carbohydrate molecules can be formed by substituting chemical groups on the basic sugar structure, and by **stereoisomerism**, that is, by varying the spatial relationship of -OH groups in the carbon chain. The sugar units in the polysaccharide are joined by a covalent **glycosidic bond**. Different polysaccharides can be formed by varying the orientation of the glycosidic bond, varying the type of carbohydrate monomer, or by mixing two or more monomers in one polysaccharide. Important polysaccharides are **cellulose**, **glycogen**, **starch**, and **peptidoglycan**.

Lipids are components of membranes, the permeability barrier of cells. Their chemical nature suits them for this task. Their **fatty acid** building blocks contain hydrophobic regions that prevent charged molecules from passing through the center of the membrane.

There are two types of **nucleic acid** (DNA and RNA). Both are composed of nucleotide building blocks. Some **nucleotides** also have other functions in energy (ATP) metabolism. All nucleotides contain a sugar, phosphate, and a nitrogen-containing base. The two nucleic acids differ in the sugar found in their building blocks. In both, polymers are formed by covalent bonds between the sugar and phosphate groups between adjacent nucleotides. This **sugar-phosphate backbone** is common to all nucleic acid molecules. The unique nature of nucleic acid molecules resides in the sequence of bases in the backbone. Four bases can occur in each nucleic acid. Three bases are found in both RNA and DNA: adenine, guanine, and cytosine. The fourth base in RNA is uracil, and thymine in DNA.

DNA consists of two sugar-phosphate backbones, which are held together by hydrogen bonds between the bases on the two **strands**. There is a preferred **base pairing** for hydrogen bond formation. The four bases belong to two different chemical classes -- **purine** or **pyrimidine**. A purine on one strand pairs with a pyrimidine on the other as follows: adenine with thymine, and guanine with cytosine. Therefore, the two strands of DNA have a **complementary** sequence. If you know the sequence of one strand, you can predict the sequence of the other by following the base pairing rules. This provides the cell with a convenient way to accurately copy the DNA sequence, which is the genetic information.

RNA differs from DNA in that it is usually single-stranded. In addition, RNA molecules are at most a few thousand nucleotides long, whereas DNA molecules may contain millions of nucleotides. There are three classes of RNA, which are important in converting the DNA nucleotide sequence in a gene to a sequence of amino acids in a protein.

Proteins consist of chains of **amino acids** joined by covalent **peptide bonds**. The bond forms between the characteristic **amino group** of one amino acid and the **carboxylic acid** group of another. The 20 amino acids differ in the chemical properties of their **side chains**. The characteristics (for example, hydrophobicity or hydrophilicity) of parts of protein molecules are determined by the types of amino acids in that part of the molecule. The diverse properties of amino acids mean that an unlimited number of different proteins can be created by varying the amino acid sequence.

Proteins do not exist as linear polymers. Rather, they are folded in various ways. Biochemists define four levels of protein structure. The **primary structure** is merely the sequence of amino acids. **Secondary structure** refers to the formation of a helix or sheet by the polypeptide chain as a result of hydrogen bonds between atoms. More extensive folding that is stabilized by non-covalent or covalent (-SH) bonds is termed **tertiary structure**. Finally, several polypeptides may associate to form a functional protein. These arrangements are the **quaternary** (4th level) **structure**. These folding properties are essential in giving the protein a specific shape to which other molecules can bind by way of non-covalent chemical interactions. This can be demonstrated by **denaturing** a protein, and observing that it then loses its biological function in catalyzing a chemical reaction or forming a cell structure.

When four different moieties are bound to a carbon atom, such molecules may exist as **stereoisomers**. Such molecules have the same structural formulas, but one is the mirror image of the other, just as the left hand is a mirror image of the right hand. With sugars and amino acids, such forms are given the designations D and L. D sugars predominate in biological systems, whereas L forms predominate in amino acids.

SELF TESTS

COMPLETION

1. The six most numerous chemical elements in cells are _____.

2. All molecules, whether in cells or in non-living matter, are composed of _____.

3. The atomic number of an element is equal to the number of _____ it contains.

4. The strongest type of chemical bonds are _____ ones.

5. The two strands of a DNA molecule are held together by _____.

6. In general, _____ are the most hydrophobic macromolecules in cells.

7. _____ is the macromolecule which is most important in allowing cells to replicate themselves.

8. Carbohydrates have a characteristic ratio of C, H, and O atoms of ___:___:___.

9. Simple lipids are composed of one _____ molecule and three _____.

10. The carriers of chemical energy in the cell are often _____, especially _____.

11. The base pairing between the two strands of DNA always involves interactions between a _____ and a _____.

12. The amino acid cysteine contains a _____ atom which may be important in stabilizing the tertiary structure of a protein.

13. Covalent bonds are _____ stronger than hydrogen bonds because covalent bonds share a pair of _____.

14. Hydrophobic interactions occur, because hydrophobic materials tend to _____ together. Hydrophophic interactions play an important role in the binding of substrates to _____.

15. _____ are the building blocks of lipids.

16. Fatty acids are linked to glycerol by _____ linkages.

KEY WORDS AND PHRASES

adenosine triphosphate (ATP) (p 33)	alpha helix (p 48)
amino acid (p 46)	atom (p 33)
atomic nucleus (p 34)	atomic number (p 34)
atomic weight (p 34)	beta sheet (p 48)
carboxylic acid group (p 46)	covalent bond (p 35)
denaturation (p 49)	deoxyribose (p 43)
disulfide linkage (p 46)	electron (p 33)
glycosidic bond (p 29)	hydrogen bond (p 35)
hydrophilic (p 36)	hydrophobic (p 36)
isomers (p 50)	isotope (p 34)
lipids (p 40)	macromolecule (p 39)
molecule (p 34)	monomer (39)
nonpolar (p 39)	peptide bond (p 47)
phosphodiester bond (p 44)	polarity (p 38)
racemase (p. 51	radioisotope (p 34)
ribose (p 32)	stereoisomer (p 51)
tertiary and quaternary structure (p 49)	triglycerides (p 40)

MULTIPLE CHOICE

1. Very gentle denaturing of a protein with urea, will **permanently** destroy what level of protein structure:

    (A) primary   (D) quaternary

    (B) secondary   (E) none

    (C) tertiary

2. The electrons in atoms are located

    (A) in the atomic nucleus

    (B) next to their associated neutron

    (C) in "shells" orbiting the nucleus

    (D) in hydrophobic portions of the atom

    (E) none of the above

3. Molecules that are attracted to water are

    (A) hydrophilic   (D) nonpolar

    (B) hydrophobic   (E) macromolecules

    (C) soluble in lipid

4. What proportion of the weight of microbial biomass is water?

    (A) 10%   (D) 80%

    (B) 30%   (E) 95%

    (C) 50%

5. Radioisotopes differ from other isotopic forms of an atom in that they :

    (A) are heavier   (D) have excess neutrons

    (B) are lighter   (E) have excess electrons

    (C) have unstable nuclei

6. Hydrogen bonding is important in which of the following?

    (A) the proper folding of polypeptide chains

    (B) the formation of double-stranded nucleic acids

    (C) the formation of peptide bonds

    (D) A and B above

    (E) all of the above

7. Which of the following may be found in complex lipids, but are never found in simple lipids?

    (A) glycerol

    (B) phosphate

    (C) fatty acids

    (D) sulfur

    (E) more than one of the above

8. The tertiary structure that a protein will achieve is a function of which of the following?

    (A) the secondary structure

    (B) the primary structure

    (C) its solubility

    (D) A and B above

    (E) all of the above

9. The quaternary structure of a protein reflects?

    (A) the primary, secondary and tertiary structure of the molecule

    (B) only the primary structure

    (C) only the tertiary structure

    (D) the primary, secondary and tertiary structure of each subunit

    (E) the number of cysteine subunits

10. Harsh denaturation (100°C) of a protein will alter?

    (A) the primary, secondary, tertiary and quaternary structure of the molecule

    (B) only the primary structure

    (C) only the tertiary structure

    (D) the covalent bonds

    (E) the secondary, tertiary and quaternary structure of each subunit

11. Pyrimidine bases go into the formation of which of the following?

    (A) Cytosine   (E) Uracil

    (B) Adenine   (F) Thymine

    (C) Guanine   (G) A, E, and F above

    (D) B and C above   (H) A and F above

12. The best example of an amphipathic material is?

    (A) ribose

    (B) lipid

    (C) nucleic acid

    (D) polysaccharide

    (E) more than one of the above

13. When studying double-stranded DNA, knowing the molar amount of guanine will also allow you to know?

    (A) the molar amount of thymine

    (B) the molar amount of cytosine

    (C) the molar amount of adenine

    (D) the stability of the molecule

    (E) more than one of the above

14. The information content of DNA is determined by?

    (A) the molar amount of thymine

    (B) the molar amount of cytosine

    (C) the sequence of bases along the polynuclear chain

    (D) translation at the surface of the ribosome

    (E) the antiparallel structure of DNA

15. Which of the following is the proper ranking (low to high) for chemical bond strength?

   (A) covalent bonds > non-covalent bonds

   (B) Van der waals interactions = Hydrogen bonds < single < double < triple

   (C) Van der waals interactions = Hydrogen bonds > single > double > triple

   (D) all bond strengths are the same

   (E) A and B above

   (F) A and C above

MATCHING

Match the chemical structure or name, with the proper species.

Structure or name	Species
1. cytosine	(A) triple bond
2. adenine	(B) keto group
3. uracil	(C) aldehyde group
4. carbohydrates	(D) ether group
5. guanine	(E) ester group
6. thymine	(F) alcohol group

7. $\overset{\overset{O}{\|}}{C}$-OH   (G) amino group

8. -NH_3 (H) carboxylic group

9. -C-OH   (I) sugar

10. CH_2-O-$\overset{\overset{O}{\|}}{C}$ (J) purine base

11. $\overset{\overset{O}{\|}}{C}$-H   (K) pyrimidine base

12. -$\overset{\overset{O}{\|}}{C}$ -

13. CH_2-O —CH_2

FILL IN (Identify the following structures)

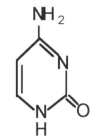
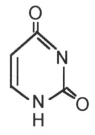

1. _____   2. _____   3. _____

DISCUSSION

1. This chapter described a series of components of matter, ranging from sub-atomic particles to macromolecules. Build up to the complex level of macromolecule structure by citing the constituents of each level of matter, beginning with the structure of the atom.

2. What chemical bonds and atomic interactions are important in biological chemistry? For each type, how do electrons contribute to the formation of the bond or interaction?

3. What are the bonding characteristics of carbon that make it a suitable element to form the basis of biological molecules?

4. Why is water an important biological solvent?

5. List the four groups of macromolecules found in cells. For each group, (a) what building block is used to construct the polymer, and (b) how are the building blocks covalently linked to form the polymer?

6. The six major chemical elements in cells are C,H,O,N,P and S. For each macromolecule, list which of these six are components.

7. Construct a table that compares RNA and DNA with respect to their constituent sugars and bases, and their cellular function.

8. The linear sequence of amino acids in a protein is called its primary structure. However, the protein's biological activity depends upon its proper folding. What types of molecular interactions are characterized by secondary, tertiary, and quaternary structure that lead to a functional protein?

9. Why are proteins and nucleic acids considered to be "informational" macromolecules? Where is the information?

10. What is the relationship between compound A and compound B shown below?

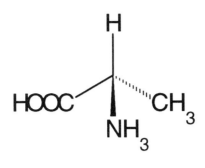

COMPOUND A   COMPOUND B

ANSWERS

Completion

1. C,H,O,N,P,S; 2. atoms; 3. protons; 4. covalent; 5. hydrogen bonds; 6. lipids; 7. DNA; 8. 1,2,1; 9. glycerol, fatty acids; 10. nucleotides, ATP; 11. purine, pyrimidine; 12. sulfur; 13. electrons; 14. cluster, enzymes; 15. fatty acids; 16. ester linkages.

Multiple Choice

1. ; 2. C; 3. A; 4. D; 5. C; 6. D; 7. D; 8. D; 9. D; 10. E ; 11. G; 12. B; 13.B; 14.C; 15.E.

Matching

1. K; 2. J; 3. K; 4. I; 5. K; 6. K; 7. H; 8. G; 9. F; 10. E; 11. C; 12. B; 13. D.

Fill In

1. Cytosine; 2. Thymine; 3. Uracil

Discussion

1. See text sections 2.1 and 2.4.

2. See text section 2.2.

3. See text section 2.2.

4. See text section 2.3

5. See text sections 2.5 - 2.8.

6. See text section 2.5 - 2.8.

7. See text section 2.7.

8. See text section 2.8

9. See text section 2.4

10. See text section 2.9

CELL BIOLOGY

OVERVIEW

Chapter 3 (pages 54-110) covers the structure and function relationships in the microbial cell. The actions occurring in the prokaryotes are contrasted with actions occurring in the eukaryotes.

CHAPTER NOTES

Microbial cells are much too small to been seen directly. Therefore, to study the morphology of cells, microscopes are used. Light microscopes are used routinely to determine cell morphology. A variety of optical techniques (**bright field**, **phase contrast fluorescence**) have been developed for specific applications, and stains can be used to highlight and increase the contrast of particular cell structures. The limit of resolution for the light microscope is about 0.2 μm. The most important staining technique is the **Gram stain**, because it differentiates organisms on the basis of cell wall structure. It results in two groups of bacteria: Gram-positive and Gram-negative. Fine details of intracellular structure can be resolved with the **electron microscope**.

Both Bacteria and Archaea cells can be divided into a number of structures that perform particular functions. It is important to know the function of these structures and their chemical composition. Some essential structures are found in all prokaryotic and eukaryotic cells. These include the **cytoplasmic membrane**, **ribosomes**, and the **genome**. Several shapes of bacteria can be recognized: spherical shaped (coccus); cylindrical shaped (rod); and curved rods (spirilla). In contrast to prokaryotic cells, eukaryotic cells are larger and much more complex. Their nucleus is contained inside a membrane and is organized into distinct DNA molecules called chromosomes. The eukaryotic cell also contains secondary structures called organelles. These organelles include mitochondria and chloroplasts (in photosynthetic organisms).

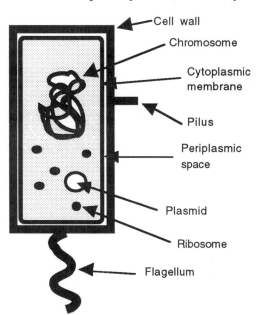

The small size of microbes has several implications. Bacterial size ranges from 0.1-0.2 μm in width to greater than 50 μm in diameter. In general, small cells grow faster than large cells. This is a reflection of their large surface to volume ratio that allows a rapid exchange with the external environment. The reason is that the nutrients to support metabolism in the **cytoplasm** must be transported across the cell membrane. The surface to volume ratio is a measure of how much membrane surface area is available to supply nutrients for the proteins in a given volume of cytoplasm. As cell size is increased, volume increases at a faster rate than surface area.

The **cytoplasmic membrane** is a selective barrier separating the cytoplasm from the external environment. In chapter 1, the cell was described as an ordered structure. To maintain order, there must be control over what enters and exits the cell. This is the function of the cell membrane.

The barrier to uncontrolled passage across the membrane is the **phospholipid bilayer**. The hydrophilic glycerol groups of lipids are aligned at the periphery of the membrane, whereas the hydrophobic fatty acids are buried in the membrane interior. The hydrophobic interior prevents passage of molecules that have a positive or negative charge. These molecules can only pass the membrane by way of specific **transmembrane proteins** that are embedded in and span the lipid matrix. The important point is that the cell can regulate the types and activities of transport proteins to control the movement of molecules into and out of the cell.

Although all cells have lipid-containing plasma membranes, the type of lipid can differ; eukaryotes, sterols are found in the membrane. A hallmark of Archaea, as compared to Bacteria or Eukarya are the **ether linkages** between the glycerol and the hydrophobic side chains contained in their lipids, rather than the **ester linkage** found in other organisms. As a result, archaeal lipids have side chains composed of the branched chain isoprene.

The lipid membrane is a barrier to the passage of molecules into and out of the cell. It is the **cytoplasmic membrane proteins** that are responsible for metabolic activities associated with membranes. The outside face of the membrane is associated with nutrient transport, whereas membrane-associated electron transport is initiated at the inside face. Those proteins that are embedded in the lipid matrix must have an external surface that is hydrophobic to enter into the membrane.

Many membrane proteins function in the transport of specific compounds, by binding them and facilitating their passage across the nonpolar membrane interior. These are referred to as **membrane transport proteins**. **Uniporters** move substance from one side of the membrane to the other. **Symporters and antiporters** move both a substance of interest and a second molecule across the membrane. Symporters carry both substances in the same direction; antiporters carry one substance in and, at the same time, one substance out of the cell. If the nutrient concentration in the environment is low, transport systems can concentrate nutrients in the cell; this requires the expenditure of energy. Transport proteins are similar to enzymes in that a particular protein may bind only one specific molecule, or a small group of chemically related compounds. **Ionophores** are compounds that destroy the selective properties of the membrane.

Facilitated diffusion is a carrier-mediated type of diffusion, which does not involve the expenditure of energy. Compounds may be transported in an unaltered form (**active transport**), or may be chemically modified (**group translocation**). The **phosphotransferase system** is an example of the latter. Sugars become **phosphorylated** during transport; phosphoenolpyruvate is the ultimate phosphate donor. Because the phosphate bond is a high energy one (as in ATP), energy is released during this process. .In active transport an energy-dependent pump is developed and the substance is moved across the membrane in an unaltered form. In active transport, the energy that drives transport is a membrane potential, created by generating a gradient of H^+ ions across the membrane. As substrates are transported by specific membrane protein carriers, protons also move across the membrane into the cell and the membrane potential decreases. Proton gradients are discussed in more detail in Chapter 4 (p 104).

Almost all prokaryotes require a **cell wall** to protect them from the effects of changes in turgor pressure. Prokaryotic walls contain unique chemical compounds not found in eukaryotes. Bacteria with a thick **peptidoglycan** (or murein) layer, and no other major wall structures stain **Gram positive**. Cells that stain **Gram negative** have a thin layer of peptidoglycan, and an **outer membrane** as part of the cell wall.

The peptidoglycan layer is a sheet (or sheets) of polysaccharide, formed by cross-linking **glycan** chains with peptide bonds between amino acids associated with different chains. Cross-linking increases the strength of the rigid layer. The repeating unit of the peptidoglycan polymer is composed of two sugar derivatives (N-acetylglucosamine and N-acetylmuramic acid) plus a small chain of amino acids. Peptidoglycan is found in all bacteria (except mycoplasmas which lack cell walls), but does not occur in Archaea or eukaryotes. However, certain methanogenic Archaea contain a cell wall composed of a polysaccharide that is similar to peptidoglycan. This material is called pseudopeptidoglycan.

The outer membrane of Gram negative bacteria contains lipid covalently linked to polysaccharide (**lipopolysaccharide**) in addition to phospholipid. This substance is important in pathogenic bacteria because it is (a) toxic to animals and (b) rather variable (in the **O-polysaccharide**) among strains. The variability allows strains to evade the body's immune system.

The outer membrane is more permeable than the plasma membrane. It contains **porins**, proteins that form small holes in the membrane. However, macromolecules cannot fit through these holes. Thus, proteins excreted through the cytoplasmic membrane are trapped in the **periplasmic space** between the inner and outer membranes in Gram negative bacteria. These proteins include **hydrolytic enzymes**(the first step in food degradation), **binding proteins** (necessary for nutrient transport) and **chemoreceptors** (used in chemotaxis response).

Why do most bacteria require a cell wall? The concentration of solutes is generally much higher inside the cell than outside the plasma membrane. The inequality can be corrected by the passage of water through the membrane into the cell (**osmosis**). However, so much water would come in that the cell would burst. The mechanical strength of the cell wall resists the osmotic pressure. This can be proven by hydrolyzing the peptidoglycan with the enzyme **lysozyme**. Unless the cells are osmotically stabilized by placing them in a concentration of sucrose that is in balance with what is inside the membrane, the cells swell and **lyse**.

New peptidoglycan is formed by synthesizing the monomers in the cytoplasm, using UDP and a lipid carrier, **bactoprenol**, to transport the monomers across the hydrophobic cell membrane, and inserting the monomer at openings created by **autolysins**. The tetrapeptides on adjacent glycan chains are then cross-linked. This **transpeptidation** is the site of action of the antibiotic **penicillin**. In effect, penicillin prevents the strengthening of new wall material, and osmotic lysis eventually kills the cell. Since new wall is only synthesized when cells grow, penicillin will only kill growing cells.

Fungi and algae often have cell walls of polysaccharides such as cellulose or chitin. Some algae have inorganic compounds such as calcium carbonate or silica in their walls.

Both prokaryotes and eukaryotes have structures used for motility. However, their structure and mechanism of action are quite different in the two cell types. Prokaryotic **flagella** are simple structures that act as propellers by rotating. In eukaryotes, flagella and **cilia** are structurally complex and drive cells forward by a whip-like motion imparted by the sliding of **microtubules** in a fashion similar to muscle filaments.

Bacterial flagella are composed of subunits of a single protein, **flagellin**. Purified flagellin subunits can aggregate to form flagellar filaments in a test tube; thus, by way of the interactions of quaternary protein structure, flagellin can **self-assemble** in the absence of other proteins. Motion is imparted by the **basal body**, proteins imbedded in the cell wall and membrane.

Motion would be of little value to the bacterium unless it was directed. Many motile bacteria are **chemotactic**. How can these simple cells detect chemical gradients and respond to them? Bacteria have 2 behaviors: (a) swimming in a straight line (**run**) and (b) stop swimming and choose a new direction at random (**tumble**). These behaviors are determined by the direction that the flagella rotate (counterclockwise vs. clockwise). The frequency of twiddling changes as cells sense themselves moving toward or away from chemical attractants. The sensors are periplasmic proteins called **chemoreceptors**. When they bind their specific attractant, they interact with membrane proteins termed **MCP's**. The MCP's can signal a phosphorylated protein, **Che Y**, which controls the direction of flagellar rotation. In addition, the methylation of the MCP's act as a "memory". Cells respond to changes in concentration with time, not in space. The interaction of chemoreceptor with MCP induces **methylation** of the MCP protein. As the level of methylation increases, higher concentrations of attractant are necessary to maintain the run. If the attractant is removed, the methyl groups are removed from the MCP's.

There are several surface structures that are important in **attachment** of bacteria. **Fimbriae** and **pili** are made up of identical protein subunits that self-assemble (like flagella). **Glycocalyx** is a general term for extracellular polysaccharide. Some pathogens initiate infections by attaching to host tissues by way of the glycocalyx, or the structures may allow pathogens to evade host defense systems.

There are several intracellular structures that are storage polymers of carbon (glycogen and poly-β-alkanoic acid) or phosphate (polyphosphate). Elemental sulfur granules are found in sulfur-oxidizing bacteria. **Gas vesicles** are intracellular protein shells that make hollow spaces in cells, and thus can provide buoyancy.

Bacterial **endospores** are very important in applications that call for sterilization, because they are extremely difficult to kill. Only a few genera produce them. When nutrients become exhausted, the **vegetative** cell differentiates to produce one endospore. A unique compound, **dipicolinic acid**, and calcium ions may be responsible for the endospore's heat resistance. Endospores can remain dormant for decades -- even thousands of years, but **germinate** to reform a vegetative cell when nutrients are provided.

The double stranded DNA is present as a naked DNA molecule and is found as a covalently closed circular molecule that is folded and twisted on itself within the cytoplasm. This arrangement is referred to as the bacterial chromosome or a **nucleoid**. The twisting and folding of the DNA are significant: as an average *E. coli* contains 1 millimeter length of DNA in a 2-3 micrometer space. As a result of the space limitations, the cell **supercoils** the DNA so that it will fit. In general, a prokaryotic cell contains only one copy of the nucleoid. However, when the cell is undergoing growth it may contain partial copies. DNA may be transferred within a population by a number of processes: (1) **conjugation**, where the DNA is transferred during cell-to-cell contact; (2) **transduction**, where viruses transfer microbial DNA; and (3) **transformation**, where DNA free in the environment, is taken up.

Three important organelles characteristic of eukaryotic cells are mitochondria, chloroplasts, and the nucleus. All three are bound by lipid membranes, but in each case the membrane is more permeable to molecules than is the plasma membrane.

The **nucleus** contains the genetic material in eukaryotes. The genome is contained on a number of DNA molecules called **chromosomes**. RNA synthesis occurs in the nucleus, and non coding regions are excised before the messenger RNA passes through the porous nuclear membrane to the cytoplasm, where it is translated into protein on the ribosomes. In contrast, the genome of prokaryotes is on one DNA molecule, and messenger RNA does not contain non coding regions. Because the DNA is not in an organelle, the processes of transcription and translation are tightly coupled in prokaryotes.

Mitochondria are the sites of energy generation. Low molecular weight compounds can easily traverse the mitochondrial membrane. There are also internal membranes (**cristae**), where the proteins involved in energy generation are located. The **matrix** of the mitochondrion contains enzymes that metabolize organic compounds. Chloroplasts also contain internal membranes (**thylakoids**), which contain the pigments and proteins necessary to carry out photosynthetic energy capture. The energy and reducing power generated by photosynthesis is used in the **stroma** to convert CO_2 into organic carbon. Mitochondria and chloroplasts are about the size of prokaryotic cells. Section 3.15 cites evidence that these organelles arose when prokaryotic cells invaded larger cells and became **endosymbionts**.

The easiest way to summarize the differences between prokaryotic and eukaryotic cells is to say that prokaryotes are structurally simple, whereas eukaryotes are relatively complex. Table 3.5

summarizes these differences. Remember that despite these differences, the two types of cells are composed of the same chemicals; it is the assembly of chemicals that is different.

SELF TESTS

COMPLETION

1. The shape of a bacterial cell is determined by the _____.

2. In general, small cells will grow _____ than large cells.

3. The osmotically fragile form of a bacterial cell which lacks a cell wall is called a _____.

4. Gram-positive membranes differ from those of Gram-negative cells in that they do not have _____ but do have _____.

5. _____ molecules do not readily pass the plasma membrane.

6. Transport proteins that carry two substances in opposite directions across the cell membrane are called _____.

7. Active transport processes can be distinguished from facilitated diffusion because the former require _____.

8. The rigid component found in bacterial cell walls is _____.

9. The lipopolysaccharide layer of Gram negative bacteria consists of _____, _____, and _____.

10. Gram _____ walls are thicker than Gram _____ walls.

11. A bacterium which has a number of flagella surrounding the cell has _____ flagellation.

12. In chemotaxis, attractants and repellents are sensed by _____ located in the _____.

13. Bacteria may store carbon and energy intracellularly as granules of _____ or _____.

14. Endospores are formed by vegetative cells when _____.

15. The genetic material of eukaryotic cells is contained in structures called _____, which consist of _____ and _____.

KEY WORDS AND PHRASES

attractant (p 89)	*Bacillus* (p 62)
bright field microscope (p 56)	capsule (p 90)
chemotaxis (p 89)	coccus (p 61)
diaminopimelic acid (p 74)	differential stain (p 58)
diploid (p 103)	flagellar hook (p 86)
gamete (p 104)	glycan tetrapeptide (p 74)
haploid (p 103)	histone (p 103)
hydrophobic region (p 65)	lipid A (p 78)

lipoprotein (p 78)	lysozyme (p 76)
nucleosome (p 104)	peritrichous flagellation (p 86)
phase contrast microscope (p 57)	phospholipid (p 64)
phospholipid bilayer (p 64)	polar flagellation (p 85)
protoplast (p 76)	resolving power (p 56)
rod (p 61)	run (p 89)
self-assembly (p 88)	slime layer (p 90)
spirillum (p 61)	spore coat (p 99)
spore cortex (p 99)	*Staphylococcus* (p 62)
sterol (p 66)	*Streptococcus* (p 62)
surface to volume ratio (p 64)	symporter (p 69)
teichoic acids (p 76)	temporal gradient (p 89)
thin section (p 59)	tumble (p 89)
uniporter (p 69)	zygote (p 104

MULTIPLE CHOICE

1. Features unique to eukaryotes are

    1. membrane-bound nucleus

    2. microtubules in their flagella

    3. have more than one chromosome

    4. 80S ribosomal structure

    5. ability to carry out photosynthesis

    (A) 1,4; (B) 2,5; (C) 4; (D) 3,5; (E) 1,2,3,4

2. Endospores are generally resistant to which of the following?

    (A) Heating   (D) two of the above

    (B) Drying   (E) all of the above

    (C) Radiation

3. The function of the bacterial cell wall is

    (A) to prevent lysis of cells in dilute solutions

    (B) to regulate the uptake of nutrients

    (C) attachment to surfaces

    (D) as a recognition site for phagocytosis

    (E) to generate a membrane potential

4. The most distinctive difference between Gram-negative and Gram-positive bacteria in electron micrographs is

   (A) the Gram + wall is thinner

   (B) the Gram + wall has an outer membrane

   (C) the Gram - wall appears as a single band

   (D) the Gram - wall contains no peptidoglycan layer

   (E) the Gram - wall is more complex

5. Arrange the following terms in the order they would be found if you traveled from the cytoplasm to the external medium of a Gram negative cell possessing a capsule:

   1. periplasmic space

   2. phospholipid bilayer of the outer membrane

   3. O-polysaccharide

   4. cytoplasmic membrane

   5. capsule

   (A) 4,3,1,2,5; (B) 4,1,2,3,5; (C) 5,2,3,1,4; (D) 5,3,2,1,4

6. The bacterial capsule

   1. may be antigenic

   2. is required for cell viability

   3. is involved in motility

   4. may contribute to a bacterium's pathogenicity

   5. determines the cell's shape

   (A) 1,4; (B) 2,4,5; (C) 1,5; (D) 4 (E) All of the above

7. Which of the following statements about bacterial flagella is **incorrect**?

   (A) The filaments are composed of protein subunits called flagellin.

   (B) All motile bacteria have flagella.

   (C) The flagella move cells by acting as propellers.

   (D) The locomotion caused by flagella can be directed to move cells toward a source of nutrients.

   (E) Flagella increase in length by adding new material to the tip not the base.

8. Penicillin and lysozyme are similar in that

   (A) both are proteins

   (B) both are antibiotics

   (C) both are brand names of the same compound

   (D) both affect cell walls

   (E) both affect the glycocalyx

9. Bacterial pili may enhance virulence of bacterial pathogens by

    (A) increasing bacterial surface area

    (B) providing a means of attachment

    (C) being an endotoxin

    (D) transporting nutrients

    (E) none of the above

10. The main chemical component of a bacterial flagellum is

    (A) ATP

    (B) peptidoglycan

    (C) a protein called flagellin

    (D) a polysaccharide called flagellin

    (E) polyphosphate

11. One characteristic shared by endospores and vegetative cells is

    (A) heat resistance   (D) low metabolic activity

    (B) resistance to drying   (E) radiation resistance

    (C) presence of DNA

12. Cell wall synthesis during cell growth involves insertion of what material into the existing wall material.

    (A) DNA   (D) lysozyme

    (B) DAP   (E) bactoprenol

    (C) glycan

13. Prokaryote respond to chemical stimulation by changing their patterns of movement. In the presence of a chemical stimulant:

    (A) runs are longer and tumbles shorter

    (B) runs equals tumbles

    (C) tumbles are longer and runs shorter

    (D) no runs are seen

    (E) tumbles are enhanced

14. Magnetosomes, glycogen, polyphosphate and poly-β-hydoxybutyric acid are all types of:

    (A) cell wall inclusions

    (B) cytoplasmic inclusions

    (C) storage products

(D) A and B above

(E) B and C above

15. Transduction differs from transformation and conjugation in that:

    (A) free DNA is needed

    (B) virus act as vectors for DNA transfer

    (C) DNA transfer occurs from cell-to-cell contact

    (D) both free DNA and cell-to-cell contact are needed

FILL IN (Identify the parts of the following structures)

Cell walls

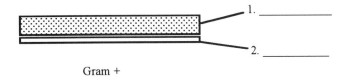

Gram +

1. _____

2. _____

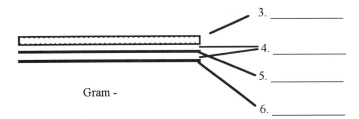

Gram -

3. _____

4. _____

5. _____

6. _____

Phospholipid bilayer

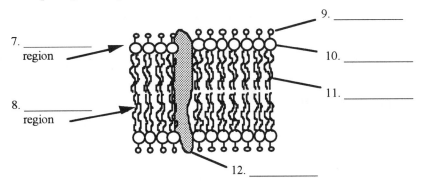

7. _____ region

8. _____ region

9. _____

10. _____

11. _____

12. _____

DISCUSSION

1. What chemical reaction in cell wall biosynthesis does penicillin inhibit?

2. Compare the chemical composition of the cell walls of Bacteria, algae, and fungi.

3. What is the function of the bacterial cell wall?

4. What is the chemical structure of peptidoglycan?

5. What is the function of the outer membrane of Gram negative bacteria?

6. How are lipids arranged in the cell membrane? Do membrane proteins tend to have high contents of hydrophobic or hydrophilic amino acids?

7. Why can't most molecules readily penetrate the cytoplasmic membrane?

8. Compare and contrast the chemical composition, structure, and function of bacterial flagella, and pili.

9. What structures are found in the basal region of the flagellum? What is the function of these structures?

10. To what environmental factors which kill vegetative cells are endospores resistant?

11. How do the mechanisms by which prokaryotic and eukaryotic flagella propel cells differ?

12. Explain how the methylation and demethylation of the MCP proteins provides a "memory" of attractant concentration in chemotaxis.

13. Eukaryotic cells can be distinguished from prokaryotes by the presence of organelles. What is the function of these three organelles: nucleus, mitochondrion, and chloroplast?

14. Compare the permeability of the plasma membrane to those of the mitochondrion and the nucleus.

15. List five structural differences between prokaryotic and eukaryotic cells.

16. Draw graphs which illustrate how the cellular rate of nutrient uptake is affected by the external concentration of nutrient if uptake is via (a) diffusion or (b) active transport. What affect would the presence of an ionophore have on the active transport process?

ANSWERS

Completion

1. cell wall; 2. faster; 3. protoplast; 4. outer membranes, teichoic acid; 5. charged; 6. antiporters; 7. energy; 8. peptidoglycan; 9. lipid A, core polysaccharide, O-polysaccharide; 10. positive, negative; 11. peritrichous; 12. chemoreceptors, periplasmic space; 13. poly-β-hydroxybutyrate, glycogen; 14. nutrients are exhausted; 15. chromosomes, DNA, protein.

Multiple choice

1. E; 2. E; 3. A; 4. E; 5. B; 6. A; 7. B; 8. D; 9. B; 10. C; 11. C; 12. C; 13. A; 14. E; 15. B.

Fill in

1. peptidoglycan; 2. membrane; 3. lipopolysaccharide/protein; 4. periplasm; 5. peptidoglycan; 6. membrane; 7. hydrophilic; 8. hydrophobic; 9. phosphate; 10. glycerol; 11. fatty acid 12. protein.

Discussion

1. See text section 3.5

2. See text sections 3.16 Table 3.5

3. See text section 3.5

4. See text section 3.5

5. See text section 3.6

6. See text section 3.3

7. See text section 3.4

8. See text sections 3.9 and 3.10

9. See text section 3.9

10. See text section 3.13

11. See text section 3.9

12. See text section 3.9

13. See text sections 3.14, 3.15 and 3.16

14. See text sections 3.14 and   3.15

15. See table 3.5

16. See figure 3.22 and text section 3.4

Chapter 4
NUTRITION AND METABOLISM

OVERVIEW

Chapter 4 (pages 111-150) begins with a consideration of microbial nutrition and culture media. Next, the chapter covers the **biosynthesis** of monomers necessary for macromolecular synthesis, and some catabolic reactions of **chemoorganotrophs (heterotrophs)**.

Metabolism is the term that pertains to all the chemical reactions in a cell. We can subdivide these reactions into those which synthesize new cell material (**anabolism**) and those whose purpose is the release of energy from the chemical energy source (**catabolism**). These two categories are linked, in that catabolic reactions provide the energy necessary to drive the anabolic reactions that result in growth.

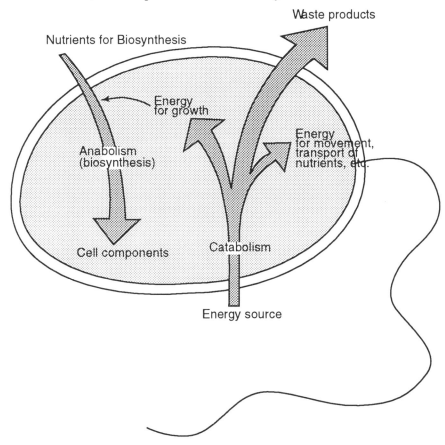

CHAPTER NOTES

Microorganisms are often divided into metabolic classes depending on their energy source. Phototrophs derive their energy from light whereas chemotrophs derive energy from chemical sources. Chemoorganotrophs use organic chemicals whereas chemolithotrophs use inorganic chemicals.

Before considering the laboratory cultivation of microbes, consider the chemical composition of cells. They are rich in protein, and the machinery to make more protein (that is, ribosomes). Thus, proteins and nucleic acids are quantitatively the most important components. In devising a **nutrient** medium for cell growth, one must consider their chemical composition. Heterotrophs require organic carbon compounds to provide C chains for biosynthesis and energy to drive biosynthesis. **Nitrogen**, most commonly provided as ammonia, is an important constituent of proteins and nucleic acids. Phosphates are commonly used as a **phosphorus** source for synthesis of nucleic acids and phospholipids. **Sulfur**, usually provided as sulfate, is essential for synthesis of two amino acids. By convention, nutrients are divided as **macronutrients** (C, H, O, N, P, S, K, Mg, Na, Ca, and Fe; Table 4.2) and **micronutrients** (Cr, Co, Cu, Mn, Mo, Ni, Se, W, V, and Zn; Table 4.3).

Nutrient media can be simple mixtures of a few pure chemical compounds, or contain a large number of preformed monomers as growth factors. As long as well-characterized chemicals are used, the medium is **defined**. However, it is simpler to add amino acids, vitamins, and other monomers as crude digests of animal or plant substances (for example, beef extract or trypticase). The medium is now **undefined**, because we do not know its precise chemical composition.

Cells have developed mechanisms to capture energy released when chemicals are **oxidized** during catabolism. Energy can be defined as the ability to do work. The amount of energy (or work) released is quantified as the change in **free energy** (ΔG). If energy is released in a reaction, ΔG (calculated as the free energy in the products minus the free energy in the reactants) is less than zero -- a negative free energy. This is true of catabolic reactions. In biosynthetic reactions, ΔG is greater than 0, and the cell must input energy to drive the reaction in the desired direction.

However, thermodynamically favorable reactions do not always go on immediately when the reactants are mixed together. Bonds must be broken to recombine molecules, and this requires **activation energy**. This could be provided by heating the reactants, but this is not feasible in cells. In cells, **enzymes** function as **catalysts** lowering the activation energy of reactions to the point that the reactions proceed at normal temperatures. The catalyst (enzyme) is unchanged by the reaction.

Enzymes are proteins that bind chemical reactants (**substrates**) at their **active site**. The three dimensional structure of the active site is complementary to the shape of the substrates. That is, they fit together just as a key fits in a lock. Chemicals differ in shape, thus enzymes are very **specific** in the substrates they bind, and thereby, the reaction(s) they catalyze. Substrate binding depends upon non covalent interactions such as hydrogen bonds and hydrophobic interactions. When substrates are bound, the activation energy necessary for the reaction to occur is lowered because the covalent bonds in the molecules are stressed.

The most important catabolic reactions are **oxidation-reduction reactions**. Although these reactions are defined by the gain or loss of electrons, in most biochemical reactions, protons are also involved. Thus, a molecule becomes **oxidized** when an electron and H^+ are removed, and it becomes **reduced** when they are added. *Therefore, oxidations do not necessarily involve oxygen*. In an oxidation-reduction reaction, electrons are donated by one compound (which becomes oxidized), and are accepted by another compound (which thereby is reduced). Therefore, we can separate the reaction into two **half reactions**: one that describes the removal of an electron from the donor, and the second that illustrates the addition of the electron to the acceptor.

Molecules differ in their tendency to donate or accept electrons. These differences are quantified as the **reduction potential** (E_0). Reduction potentials tell you what molecules will react with each other, and in which direction electrons will be transferred. The smaller (or more negative) the reduction potential, the greater the tendency to donate electrons to the oxidized form of a molecule with a more positive reduction potential. The difference in reduction potential between two half reactions is also important; it is a measure of the energy released in the overall oxidation-reduction reaction. It is the energy released from an electron donor (energy source) in coupled oxidation-reduction reactions which cells capture for use in biosynthetic reactions. The most common chemical "currency" of energy in cells is ATP; a large amount of free energy is released when either of two **high energy phosphate bonds** in the molecule are hydrolyzed.

Many oxidation-reduction reactions involve cytoplasmic enzymes. The coenzymes NAD or NADP function as **intermediate electron carriers** by accepting electrons removed from a substrate in one enzymatic reaction, and then donating the electrons to other molecules in a variety of other enzymatic reactions.

There are two types of **biochemical pathways** by which heterotrophic organisms can synthesize ATP as a consequence of the oxidation of organic compounds. These are **respiration and fermentation**. In the former, a molecule (O_2, NO_3 etc.) in the environment serves as the terminal electron acceptor. In fermentation, the organic energy source is oxidized, and an intermediate organic product is reduced to balance the overall oxidation-reduction reaction.

With fermentation, the organic energy source cannot be completely oxidized, because of the necessity for an electron donor. As a consequence, only a portion of the potential energy in its chemical bonds is released. ATP synthesis during fermentation occurs by **substrate level phosphorylation**, in which a high energy phosphate bond is donated directly by an organic compound to ADP to yield ATP. **Glycolysis (Embden-Meyerhof pathway)** is a common biochemical pathway for glucose catabolism under fermentative conditions. In this pathway, glucose is first activated by the expenditure of energy (ATP). The organic compound is then oxidized; the electrons are donated to NAD and substrate level phosphorylation occur. In most fermentations, the ATP yield is small -- 2-4 ATP per glucose. The end product of glycolysis, **pyruvate** or a metabolite derived from it, then serve as electron acceptors to regenerate the oxidized form of NAD. This last step is essential; there is only a small amount of NAD in the cell to function as an intermediate electron carrier. The nature of the **fermentation products** differs among bacterial species; many of them have commercial importance.

The energy gains in respiration are larger than in fermentations because the organic carbon can be completely oxidized to CO_2 and the amount of energy released when electrons are donated to the terminal electron acceptor is much greater. The initial biochemical steps of glucose metabolism are similar to fermentations; the glycolytic pathway is often used. However, pyruvate is metabolized in the **tricarboxylic acid cycle**; it is oxidized to CO_2 and the electrons are used to reduce NAD. **Oxaloacetate** is the acceptor molecule that reacts with a product of pyruvate metabolism (acetate), and oxaloacetate must be regenerated at the end of the cycle.

A large amount of reduced NAD is generated by these reactions, and it must be reoxidized for metabolism to continue. In respiration, this occurs in membrane-associated **electron transport systems**. These are a set of intermediate electron carriers specially arranged in the cytoplasmic membrane. The electron carriers are **flavoproteins**, **cytochromes**, and **quinones**. The specific arrangement of carriers in the membrane is such that as electrons and H^+ are passed from carrier to carrier, the protons are extruded outside the cell membrane (specifically, in reactions involving flavoproteins and quinones), while the electrons are returned to the cytoplasmic side. The result is a **proton gradient**, in which the outside face of the membrane has a lower pH and a more positive charge. This energized state of the membrane (the **proton motive force**) represents potential energy that can perform useful work. The controlled passage of protons back across the membrane through specific membrane proteins is used to drive ion transport, flagellar rotation, or ATP synthesis. This last process involves a membrane bound **ATPase**. Synthesis of ATP via generation of a proton motive force by electron transport is called **oxidative phosphorylation**.

Thus, in aerobic respiration NAD is reoxidized by an electron transport chain. Furthermore, enough energy is released in the transfer of electrons from reduced NAD to O_2 to synthesize up to 3 ATP. Theoretically, 38 ATP can be synthesized from the oxidation of one glucose molecule, and most of this is by way of oxidative phosphorylation. Note the large difference in energy gain between aerobic respiration and fermentation. However, the efficiency of the fermentation reaction can be fairly high. The rate of ATP turnover in the cell is rapid as ATP is a high energy and somewhat unstable molecule. ATP is not an energy storage material; it serves only in a catalytic capacity. For long-term energy storage, most cells will accumulate glucose polymers such as starch or glycogen as well as poly-β-hydroxybutyrate and polyhydroxyalkanoates.

Respiration does not necessarily involve O_2 as the terminal electron acceptor. In the absence of oxygen, specific bacteria may use nitrate, sulfate, or carbonate as terminal electron acceptors; these processes are termed **anaerobic respiration**. Furthermore, some bacteria do not use organic compounds as their energy source. Reduced inorganic compounds are oxidized by **chemolithotrophs** to gain energy, and **phototrophs** capture light energy and use it to generate a proton motive force and synthesize ATP.

Although microorganisms differ widely in their catabolic reactions, biosynthetic reactions (**anabolism**) are amazingly similar in diverse species. Biosynthesis requires energy (provided by catabolic reactions), and carbon skeletons to build upon. These carbon skeletons are also provided by catabolic pathways, especially by **glycolysis** and the **tricarboxylic acid cycle**.

Sugars are used for synthesis of cell wall, polysaccharide, and nucleic acids. Sugars can be obtained from the external environment or synthesized within the cell. Gluconeogenesis is the formation of glucose from non carbohydrate materials. The starting point for gluconeogenesis is phosphoenolpyruvate that is formed as part of the TCA cycle. Two key intermediates are glucose-6-PO_4 (a metabolite in glycolysis) and **UDP-glucose**, a precursor of polysaccharide biosynthesis. As a consequence, glucose-6-PO_4 is the central intermediate in glucose catabolism and UDP-glucose is the central intermediate in glucose anabolism.

The twenty amino acids can be grouped into five families, based on their precursors. The common pathway within a group illustrates how one enzymatic reaction can be used for the synthesis of more than one end product. Nitrogen is added to most amino acids by **transamination** reactions, in which an **amino group** is donated by glutamate. Cells replenish the glutamate pool by incorporating exogenous ammonia by way of the enzyme **glutamate dehydrogenase**.

Biosynthesis of the nitrogenous bases in nucleic acids is complex, and the pathways for purine and pyrimidine synthesis differ. Purines are synthesized atom by atom by adding to the sugar-phosphate portion of a nucleotide. In contrast, the pyrimidine ring (orotic acid) is synthesized before the sugar is added.

Fatty acids are the major components of lipids as well as electron donors (energy sources) for many chemoorganotrophic microorganisms. Fatty acids are synthesized by a stepwise buildup of long chains from acetyl-CoA. Figure 4.31 shows that fatty acids are built up two carbons at a time from the three carbon compound **malonate**. Addition of double bonds to **saturated fatty acids** to create **unsaturated fatty acids** may be achieved by a reaction requiring molecular oxygen (in aerobes) or by dehydration of a hydroxyl fatty acid (in anaerobes).

SELF TESTS

COMPLETION

1. An enzyme which hydrolyzed lipids would be called a _____.

2. Enzymes are composed of _____; thus, they are _____ by heat.

3. Isozymes are enzymes which catalyze the same reaction but which differ in _____.

4. Enzymes function as catalysts by lowering the _____ of a chemical reaction.

5. If the free energy change of a chemical reaction is less than 0, the reaction is _____.

6. Oxidation-reduction reactions in cells generally involve the loss or gain of _____.

7. The oxidation-reduction pairs X/XH_2 and Y/YH_2 have reduction potentials of -80 and +80 millivolts, respectively. This means that electrons would most likely be removed from ____ to reduce _____.

8. The most important high-energy compound in cells is _____.

9. The organic end product of the glycolytic pathway is _____.

10. There is net synthesis of ___ ATP and ___ reduced NAD in glycolysis.

11. If there is no exogenous terminal electron acceptor available to cells growing on glucose, then electrons in reduced NAD are donated to _____.

12. In the TCA cycle, _____ is the molecule which accepts organic carbon metabolized in glycolysis.

13. The synthesis of ATP driven by the functioning of electron transport chains is called _____.

14. The actual chemical electron acceptor/donor is _____ in flavoproteins, _____ in cytochromes, and _____ in quinones.

15. Sulfonamides can inhibit the growth of bacteria because they interfere with the synthesis of _____.

16. The energy released as electrons are passed through the carriers of the electron transport system is used to drive _____ across the membrane.

17. When the membrane potential is used to synthesize ATP, protons cross the membrane through _____.

18. The key intermediate in the biosynthesis of hexoses and polysaccharides is _____.

19. Extracellular ammonia can be incorporated into the 20 amino acids via the activities of _____ and _____.

KEY WORDS AND PHRASES

activation energy (p 121)	amino acids (p 113)
anaerobic respiration (p 140)	autotrophs (p 140)
catalyst (p 120)	chemolithotroph (p 110)
coenzyme (p 122)	complex medium (p 118)
coupled reaction (p 123)	electron acceptor (p 124)
electron donor (p 124)	electron tower (p 124)
endergonic (p 119)	exergonic (p119)
folic acid (p 146)	free energy (119
gluconeogenesis (p 143)	growth factor (p 115)
half-reaction (p 144)	heme (p 133)
heterotroph (p 140)	hexose (p 143)
intermediate electron carrier (p 125)	kilocalorie (kcal) (p 118)
kilojoule (kJ) (p 118)	macronutrients (p 115)
micronutrients (p 116)	NAD+/NADH (p 125)
oxidation (p 123)	pentose (p 144)

phosphoenolpyruvate (p 127)	phototroph (p 140)
primary electron donor (p 125)	prosthetic group (p 121)
reduction (p 123)	respiration (p 132)
saturated fatty acid (p 147)	sulfonamides (p 146)
synthetic medium (p 118)	terminal electron acceptor (p 125)
transamination (p 145)	uncouplers (p 137)
unsaturated fatty acid (p 147)	vitamin (p 116)

MATCHING

I. Match the amino acid with its precursor.

1.	leucine	(A)	glutamate
2.	tyrosine	(B)	aspartate
3.	tryptophan	(C)	pyruvate
4.	histidine	(D)	serine
5.	arginine	(E)	chorismate
6.	proline	(F)	histidinol
7.	methionine		
8.	glycine		
9.	phenylalanine		
10.	asparagine		

II. Match the prokaryotic cell component with its relative abundance (% by weight) in the cell.

1.	protein	(A)	0.5
2.	lipid	(B)	2
3.	polysaccharide	(C)	3.5
4.	DNA	(D)	20.5
5.	RNA	(E)	3.1
6.	amino acids and precursors	(F)	9.1
7.	sugars and precursors	(G)	5
8.	nucleotides and precursors	(H)	55

III. Match the element with an important cellular use.

1.	cobalt	(A)	protein synthesis
2.	zinc	(B)	cysteine formation
3.	molybdenum	(C)	stabilize ribosomes, membrane
4.	copper	(D)	enzymes in respiration
5.	manganese	(E)	carbonic anhydrase structure

6.	nickel	(F)	formation of B_{12}
7.	tungsten	(G)	molybdoflavoproteins
8.	magnesium	(H)	hydrogenases
9.	sodium	(I)	formate metabolism
10.	potassium	(J)	seawater organisms
11.	calcium	(K)	not essential, stabilizing cell walls
12.	sulfur	(L)	superoxide dismutases

MULTIPLE CHOICE

1. The inorganic form of nitrogen that most bacteria can readily use is:

    (A) nitrate

    (B) nitrite

    (C) ammonia

    (D) nitrogen gas

    (E) glutamate

2. Sulfur is needed by all bacteria. Its main use is

    (A) as an energy source

    (B) as a component of nucleic acids

    (C) in proteins

    (D) as a cell wall component

    (E) as an electron acceptor

3. Coupled reactions are found in:

    (A) energy production

    (B) catabolic reactions

    (C) anabolic reactions

    (D) oxidation-reduction reactions

    (E) all of the above

4. In aerobic respiration, the terminal electron acceptor is

    (A) sulfate

    (B) nitrate

    (C) phosphate

    (D) water

    (E) oxygen

5. Fermentation requires

    (A) light energy

    (B) the presence of organic electron acceptors

    (C) the presence of oxygen

    (D) the absence of oxygen

6. Which best describes the reasons that fermenting organisms carry out reactions which reduce pyruvate?

    (A) Pyruvate is toxic to cells.

    (B) Reduced NAD is toxic to cells.

    (C) Cells must regenerate oxidized NAD.

    (D) It is an unregulated reaction of no value.

    (E) The cell generates energy from pyruvate reduction.

7. The central intermediate leading to the production of different fermentation end products is

    (A) acetic acid   (D) lactic acid

    (B) pyruvic acid   (E) acetyl CoA

    (C) formic acid

8. The TCA cycle has two major functions. These are

    1. generation of energy

    2. production of biosynthetic intermediates

    3. production of pyruvic acid.

    4. production of fatty acids.

    5. utilization of coenzyme H.

    (A) 1,2   (B) 2,3 (C) 3,4 (D) 4,5 (E) 1,4

9. In aerobic respiration, CO_2 production primarily

    (A) takes place in the TCA cycle

    (B) occurs each time an ATP is made

    (C) occurs at each oxidative phosphorylation

    (D) occurs at each dehydrogenation

    (E) takes place in glycolysis

10. Electron transport chains are located in

    (A) ribosomes   (D) periplasmic space

    (B) the cell wall   (E) glycocalyx

    (C) membranes

11. Which of the following statements is TRUE:

    (A) Glucose is a direct precursor for the synthesis of amino acids.

    (B) Pyruvate is the direct precursor for purine biosynthesis.

    (C) Adenine and guanine are purines.

    (D) Adenine and thymine are pyrimidines.

    (E) Biosynthetic pathways are simply the reversal of degradative ones.

12. There is a core of metabolic pathways similar in all bacteria. These include:

    1. amino acid biosynthetic pathways

    2. synthesis of the peptidoglycan repeating subunit

    3. pathways for nucleotide biosynthesis

    4. ATP production by anaerobic respiration

    5. energy production in fermentations.

    (A) 1,3 (B) 2,3,4 (C) 1,3,5 (D) 3,4,5 (E) all of the above

13. An organism which carried out fermentative metabolism of glucose in the presence or absence of oxygen is considered:

    (A) an obligate aerobe

    (B) an obligate anaerobe

    (C) a facultative aerobe

    (D) microaerophilic

14. How many high-energy phosphate bonds are there in adenosine triphosphate?

    (A) 0   (C) 2

    (B) 1   (D) 3

15. What is the average protein content of a prokaryotic cell?

    (A) 5%   (C) 96%

    (B) 9.1%   (D) 55%

DISCUSSION

1. Why do cells need enzymes to mediate reactions?

2. The redox potential of a half reaction indicates the tendency of the oxidized substance to accept electrons. Do large positive redox values indicate a great or a slight tendency to accept electrons?

3. What are the most commonly used soluble intermediate electron carriers in cells?

4. Distinguish between catabolism and anabolism.

5. Compare the different types of molecules that function as electron carriers in membrane-bound electron transport systems.

6. How does a heterotrophic bacterium generate a proton motive force? How can this force be used to synthesize ATP?

7. How many ATP can theoretically be produced from the aerobic respiration of glucose via (a) substrate level phosphorylation and (b) electron transport phosphorylation? What proportion of the total energy available in glucose is captured by the cell?

8. Compare fermentations to aerobic respiration with respect to (a) the amount of energy released from the energy source, (b) the compound that serves as the terminal electron acceptor, and (c) the mechanisms of ATP synthesis that are involved.

9. In aerobic respiration, O_2 is the terminal electron acceptor, and the bulk of the ATP is synthesized via oxidative phosphorylation. Would the uncoupler dinitrophenol and the inhibitor carbon monoxide have (CO) the same effect on oxygen uptake of an obligate aerobe?

10. What are the precursor molecules of the different families of amino acid biosynthesis?

12. What is the difference between a complex medium and a chemically defined one?

13. What is the direct effect, and what secondary effects occur in obligate aerobes treated with each of these compounds: (a) uncouplers such as dinitrophenol, and (b) inhibitors of cytochromes such as cyanide?

ANSWERS

Completion

1. lipase; 2. proteins, denatured; 3. structure; 4. activation energy; 5. exergonic; 6. electrons; 7. XH_2,Y; 8. ATP; 9. pyruvate; 10. 2,2; 11. organic compounds; 12. oxaloacetate; 13. oxidative phosphorylation; 14. flavin, iron, quinone; 15. folic acid; 16. protons; 17. ATPase; 18. UDP-glucose; 19. glutamate dehydrogenase, transaminases

Multiple choice

1. C; 2. C; 3. E; 4. E; 5. B; 6. C; 7. B; 8. A; 9. A; 10. C; 11. C; 12. A; 13. C; 14. C; 15. D.

Matching

 I. 1. C; 2. E; 3. E; 4. F; 5. A; 6. A; 7. B; 8. D; 9. E; 10. B.

 II. 1. H; 2. F; 3. G; 4. E; 5. D; 6. A; 7. B. 8. A

III. 1. F; 2. E; 3. G; 4. D; 5. L; 6. H; 7. I; 8. C; 9. J; 10. A; 11. K; 12. B.

Discussion

 1. See text Section 4.5

 2. See text Section 4.6

 3. See text Section 4.7

 4. See text Section 4.1

 5. See text Sections 4.7 and 4.12

 6. See text Section 4.11

 7. See text Section 4.13

 8. See text Sections 4.8 and 4.13

 9. See text Section 4.13

10. See Figure 4.27

11. See text Section 4.3

12. See text Section 4.13.

MICROBIAL GROWTH

OVERVIEW

Chapter 5 (pages 151-179) is concerned with the biological and biochemical basis of cell growth. Microbial **growth** is defined as an increase in either cell numbers or total cell mass. Although the chapter focuses on prokaryotes, the general principles described apply to the growth of all unicellular microorganisms.

CHAPTER NOTES

These biosynthetic reactions occur in individual cells, but, because of their small size, growth studies must be conducted on **populations** of microorganisms. An increase in bacterial cell size inevitably leads to formation of new cells by **binary fission**, so growth can be followed by monitoring cell number or mass in the population.

When a cell divides into two, each new cell can catalyze the biosynthesis of additional cell material. Thus, the number of catalytic units increases with time; as a consequence, biomass is synthesized at an ever increasing rate in the growing culture. Individual cells grow at the same rate as before, but there are many more cells present to make biomass. This is **exponential growth** and is described by the equation: $N = N_O 2^n$. Where N is the population size at the final time, N_O is the initial population size and n is the number of generations. The generation time is defined by t/n where t is the elapsed time and n is the number of generations. (The time to go from one cell to two cells is called a generation time.) This change in cell numbers can be taken into account by graphically plotting the **logarithm** of cell number versus time.

The growth of a culture can be monitored by measuring changes in cell number or total biomass. Many different methods have been developed to estimate these parameters. There is no perfect method -- the choice depends upon the particular situation.

Cell number can be determined by direct microscopic counts or by viable counts. **Direct microscopic counts** are limited by the inability to distinguish living from dead cells. Furthermore, cell densities greater than 10^6 per ml are required to see enough cells to count. In contrast, **viable counts** can detect very small numbers of cells; one is limited only by how much material can be inoculated onto the surface of the agar medium. This method detects only living cells; that is, those capable of forming **colonies** on an appropriate nutrient medium. Viable counts can take the form of spread plates and pour plates. The difference is in the timing of cell entry on to the medium. The basis for the technique is that individual cells are deposited at separate locations on the plate so that each colony that arises was derived from an individual cell. If this assumption is not true, the estimate of cell number will be in error. **The assumption is violated if cells in the sample are in clumps, or if too many cells were inoculated onto the plate**. In the latter case, it becomes likely that more than one cell is placed on a spot. Therefore, the resulting colony arose from two, not one cell. On the other hand, if too few cells are plated, the statistical precision of the result is poor. Therefore, microbiologists try to plate 30 to 300 cells per plate for viable counts.

Total cell biomass can be monitored by measuring the weight of cells in the culture. In practice, water is removed by drying before measuring **dry weight**. A large number of cells are needed to obtain enough biomass to weigh accurately, as an individual bacterium weighs about 10^-

grams. A simple, rapid, and convenient method of biomass estimation is **turbidity** measurement. Cells scatter light -- the more light that is scattered, the more cells (and biomass) in the sample.

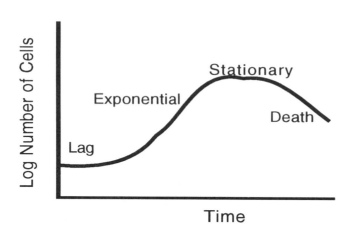

When bacteria are inoculated into a flask containing a fixed amount of nutrient medium (a **batch culture**), the growth of the culture undergoes several characteristic phases. In **lag phase**, cell metabolism is directed towards synthesizing enzymes necessary for growth in the particular medium. Thus, the lag is longest if the inoculum consisted of non-growing cells or if they came from a medium of very different composition. **Exponential phase** is the growth period, where cells undergo binary fission to logarithmically increase the population size. This explosive rate of growth cannot be maintained indefinitely in the fixed amount of medium. At some point, an essential nutrient is depleted or a toxic metabolic product accumulates to an inhibitory concentration. Growth stops, and the population size reaches a plateau, in what is called **stationary phase**. However, the cells are still metabolizing, and may begin producing **secondary metabolites**, some of which are of industrial importance.

The concentration of a nutrient may affect either **growth rate** or **growth yield** in a culture. Effects on growth rate generally occur only at very low concentrations, where the rate of nutrient uptake does not sustain the cell's demands. As mentioned above, stationary phase in a batch culture occurs if an essential nutrient is depleted. In this case, the initial concentration determined the **total crop** of cells. The amount of biomass that can be synthesized from a specific energy source is related to the amount of energy captured by the cell in catabolism. In general, 9-10 g of dry weight biomass are produced per mole of ATP generated from catabolism.

Continuous culture is an alternative means of growing microbes. It differs from a batch culture in that it is an open system, in which fresh medium is continuously added, and culture is removed so that a constant volume is maintained. In this system, cells will grow exponentially for extended periods. Furthermore, the system has the property of reaching **steady state** in which the concentration of limiting nutrient and the cell number do not vary with time.

The experimenter can control the nature of the growth-limiting nutrient by adjusting the composition of the inflow medium such that one nutrient is in relatively low concentration. The growth rate of the cells in the culture is determined by the rate at which fresh medium is pumped into the culture. The cell density at steady state is determined by the concentration of limiting nutrient in the inflow medium -- only so much biomass can be constructed

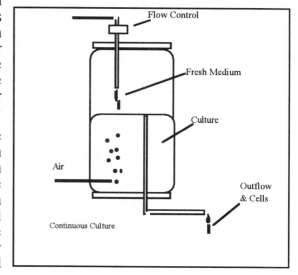

from a given amount of a nutrient. As fresh medium is pumped into the culture more rapidly, the cells grow faster because the limiting nutrient is supplied faster. However, cell density remains constant because culture is also removed faster through the overflow of the chemostat.

During exponential growth in either batch or continuous cultures, growth is **balanced**. That is, when a cell doubles to form two cells, the amount of each and every cell component is doubled. The composition of cells changes as environmental conditions change. Therefore, if a batch culture is transferred to a different medium, or the **dilution rate** of a continuous culture is changed, growth becomes unbalanced as cells adjust their macromolecular composition. Changes in the rate of ribosomal RNA synthesis are observed when these shifts are made, because growth rate is in large part a reflection of how fast cells make protein. The number of ribosomes per cell is proportional to growth rate. Ribosomal RNA constitutes 80% of total RNA in the cell, and is very stable. In contrast, messenger RNA is broken down a few minutes after it is synthesized in bacteria. Therefore, to continue production of a particular protein, there must be continual transcription of its gene.

E. coli has a generation time of 30 minutes under some conditions. However, 40 minutes are required to replicate the chromosome. This disparity is resolved by beginning a second round of DNA replication before the first is completed. In this case, the genome is already partially replicated when it is segregated into daughter cells just before cell division.

The growth of bacteria depends upon not only the provision of nutrients, but also the existence of appropriate environmental conditions such as temperature, pH, water activity, and aeration. Species differ in the range of these factors within which they will grow.

As **temperature** increases, chemical reactions can proceed faster. Thus, cells should grow faster as temperature is raised. However, there is a limit beyond which some temperature-sensitive macromolecules such as a protein, nucleic acid, or lipid will become denatured, and therefore nonfunctional. There is also a minimum temperature for growth, below which the lipid membrane is not fluid enough to function properly. Note that the optimum temperature is nearer the maximum than the minimum.

In general, organisms can grow over a temperature range of 30 to 40°C. However, species differ in range, and four categories have been delineated on this basis. **Mesophiles** have optimal growth temperatures in the range 20-50 degrees C; that is, the temperatures most common on the earth's surface or in animals. **Psychrophiles** have optimal temperatures below 15°C. These organisms are killed by exposure to room temperature. They function at low temperature by having high contents of **unsaturated** fatty acids in their membranes. These molecules remain fluid at temperatures where membranes containing saturated fatty acids are nonfunctional.

Psychrotrophs can cause spoilage of refrigerated products, such as food or blood. They grow fastest at temperatures above 20°C, and therefore are likely to contaminate these products, but are capable of slow growth at refrigerator temperatures.

Organisms that grow best above 50°C are **thermophiles**. Some bacteria can grow up to temperatures where water boils; those with optimal growth temperatures above 75°C are categorized as **extreme thermophiles.** Most of these are Archaea. No eukaryotic microbe can grow at these high temperatures, and photosynthetic microbes are not found at temperatures that support other metabolic types. Thermophiles contain proteins and lipids that are not denatured at the high temperatures where they grow.

Most microbes grow somewhere within the **pH range** of 5 to 9; most natural environments fall within this range. However, species do exist which grow at pH extremes. Fungi tend to grow at lower pH values than do bacteria.

Microbial species vary in their requirement for **water availability**. This is a function of not only water content, but also the presence of solutes such as salt or sugar, which by **osmosis** will cause water to diffuse out of the cell. Marine bacteria are optimally adapted to the salt concentration in sea water -- growth is inhibited at higher or lower concentrations, and their enzymes have a specific requirement for sodium ions. Species that grow in media with high solute concentration balance the external solute with a high concentration of internal solute. However, this compound must be one that does not inhibit the activity of cytoplasmic enzymes; it must be a **compatible solute**.

To grow microbes anaerobically, procedures to exclude air from the culture must be used. These include filling bottles completely to exclude a gas space or adding **reducing agents** to react with any oxygen gas. Obligate anaerobes are killed by exposure to oxygen gas. Therefore, all manipulations must be performed in an anaerobic atmosphere, such as an **anaerobic glove box**.

All organisms produce toxic products from oxygen when exposed to O_2. All of them except obligate anaerobes have enzymes that detoxify these compounds. Toxic products are formed by reaction of oxygen with oxidation-reduction carriers or pigments (in the presence of light). The products are highly reactive, and can destroy essential cell components such as lipids, proteins, or nucleic acids. Superoxide dismutase, catalase, and peroxidase are three enzymes that destroy specific toxic compounds.

SELF TESTS

COMPLETION

1. Bacteria are in exponential phase of growth when a straight line is obtained by plotting _____ as a function of time.

2. The amount of biomass in a culture containing more than 10^7 cells per ml can be determined most quickly and easily by _____.

3. A low nutrient concentration may limit the growth _____ of a culture; the nutrient concentration may also limit the growth _____ in a batch culture.

4. Continuous cultures differ from batch cultures in that the cells can be maintained in _____ phase of growth for extended periods of time.

5. In a continuous culture, the growth rate is equal to the _____.

6. Most of the energy a cell consumes to grow is used for _____.

7. When an exponentially growing culture is transferred to a richer medium, one of the first observable changes is an increase in the rate of _____.

8. The amount of biomass produced from an energy substrate is related to the amount of _____ generated from it.

9. The environmental trigger for bacterial endospore formation is _____.

10. As the growth temperature of an organism increases, it seems as if the proportion of _____ fatty acids in the cell membrane must increase.

11. Most microbes have pH optima in the range _____.

12. As the solute concentration (i.e. sucrose) increases, the water activity _____.

13. Osmophilic microbes balance the high external solute concentration with an internal _____.

14. Anaerobic bacteria that are killed by exposure to atmospheric oxygen generally lack the enzyme _____.

15. Organisms killed by brief exposure to oxygen are _____.

KEY WORDS AND PHRASES

acidophile (p 170)	aerotolerant anaerobe (p 174)
alkaliphile (p 171)	a_W (p 171)
batch culture (p 160)	binary fission (p 153)
cardinal temperature (p 164)	catalase (p 176)
cell density (p 162)	chemostat (p 160)
colony count (p 158)	colony forming unit (p 159)
compatible solute (p 172)	continuous culture (p 160)
death phase (p 157)	doubling time (p 153)
exponential phase (p 156)	facultative aerobe (p 175)
generation time (p 155)	growth rate (p 154)
growth yield (p 162)	halophiles (p 172)
hydroxyl free radical (p 176)	hyperthermophiles (p 164)
lag phase (p 155)	maintenance energy (p 162)
maximum growth temperature (p 164)	mesophiles (p 164)
microaerophile (p 341)	minimum growth temperature (p 164)
neutrophile (p 171)	obligate aerobe (p 341)
obligate anaerobe (p 174)	optimum growth temperature (p 164)
osmophile (p 172)	peroxidase (p 176)
peroxide (p 176)	pH (p 170)
plate count (p 158)	pour plate (p 158)
psychrophiles (p 164)	psychrotolerant (p 166)
singlet oxygen (p 176)	spread plate (p 158)
stationary phase (p 156)	*Sulfolobus* (p 170)
superoxide (p 176)	superoxide dismutatase (p 176)
thermophiles (p 164)	thermophiles (p 164)
Thiobacillus (p 170)	thioglycollate (p 175)
total cell count (p 157)	triplet oxygen (p 176)
water activity (p 171)	xerophiles (p 172)

MULTIPLE CHOICE

1. Which feature below typifies binary fission?

    (A) One cell divides in one generation to produce four cells.

    (B) The resulting daughter cells are unequal in size.

    (C) The daughter cells contain approximately equal amounts of each cell component.

    (D) The genome is not replicated until septum formation is completed.

    (E) All of the cell wall material of the new cell is newly synthesized.

2. A bacterium with a generation time of 0.5 hours will grow from a population of 10^3 to 10^9 cells in about how many hours?

    (A) 3   (D) 25

    (B) 10   (E) 40

    (C) 20

3. The generation time of a culture is defined as

    (A) the time it takes the culture to begin growth.

    (B) the interval of time between the end of log phase to the stationary phase of growth.

    (C) how long the culture has been growing.

    (D) the period of time it took for the organism to appear on earth.

    (E) the time it takes a cell to divide.

4. If bacteria in the log phase of growth were transferred to fresh medium of the same composition, which phase of a batch culture would not be observed in this new culture?

    (A) lag phase   (C) stationary phase

    (B) log phase   (D) death phase

5. Choose the TWO correct answers. A bacterial culture enters stationary phase because:

    1. the cells are "tooling up" for rapid growth

    2. toxic products of metabolism may have accumulated

    3. the energy source may have been depleted

    4. the cells have aged and old cells cease dividing

    5. it must synthesize new proteins before it can resume growth

    (A) 1,5 (B) 1,2 (C) 2,3 (D) 1,4 (E) 2,4

6. Which is the most SENSITIVE method of enumerating microbes?

    (A) viable plate count

    (B) direct microscopic count

    (C) dry weight measurement

    (D) turbidity measurement

    (E) measurement of cell protein

7. Turbidimetric methods for determining microbial biomass are

    (A) easy to do and accurate with small numbers of bacteria

    (B) difficult to carry out, but very accurate

    (C) easily and quickly done, but require relatively large numbers of cells

    (D) effective in determining only viable cells

    (E) highly specialized techniques not suitable for routine laboratory use

8. Which of the following methods of preservation reduces the water activity of foods?

    (A) curing meats with salt   (D) pickling

    (B) refrigeration   (E) canning

    (C) pasteurization

9. The growth condition in which each cell component doubles in amount when a cell doubles is:

    (A) cryptic growth   (D) balanced growth

    (B) exponential growth   (E) binary fission

    (C) maximum growth

10. The steady-state level of biomass in a continuous culture is determined by:

    (A) the concentration of limiting substrate in the culture vessel.

    (B) the concentration of limiting substrate in the medium reservoir.

    (C) the dilution rate.

    (D) the temperature.

    (E) the volume of the culture vessel.

11. Microbes with temperature optima below 20°C are called:

    (A) psychrotrophs   (D) mesophiles

    (B) psychrophiles   (E) xerophiles

    (C) thermoduric

12. Osmotolerant yeasts are able to grow at high salt concentrations because their cytoplasm contains high concentrations of

    (A) sodium ions   (D) lipid

    (B) polyalcohols like glycerol   (E) amino acids

    (C) catalase

13. The enzyme catalase detoxifies which of the following molecules?

    (A) singlet oxygen   (D) hydroxyl free radical

    (B) triplet oxygen   (E) peroxide

    (C) superoxide anion

14. Arrange the following techniques in order of increasing stringency for maintaining anaerobic microbes.

    1. addition of a reducing agent

    2. use of an anaerobic jar

    3. use of an anaerobic glove box

    4. fill bottles completely with media

    (A) 1,2,3,4 (B) 4,2,3,1 (C) 3,1,4,2 (D) 4,1,2,3 (E) 2,3,4,1

MATCHING

Match the type of microorganism with its growth condition

Type of Organism	Growth Condition
1. psychrophile	A) growth at low pH
2. mesophile	B) growth requires NaCl
3. thermophile	C) growth at mid-range temperatures
4. hyperthermophile	D) growth at high sugar concentrations
5. acidophile	F) growth near pH 7
6. neutrophile	G) growth at high pH
7. alkliphile	H) growth at high temperatures
8. halophile	I) growth in very dry conditions
9. osomophile	J) growth at low temperature
10. xerophile	K) growth at very high temperatures

DISCUSSION

1. What are the differences between the growth of individual bacterial cells and the growth of a population of bacteria?

2. Explain the following terms: generation time, exponential growth, and binary fission.

3. List the advantages and disadvantages of each of the following methods that can be used to measure microbial biomass or numbers: (a) direct microscopic count, (b) viable plate count, (c) turbidity measurement, and (d) dry weight measurement.

4. Microbial cultures growing in a closed system (a flask, a test tube, an agar plate) undergo four distinct phases of growth. What are these phases? Explain why they occur and what the cells are doing in each phase.

5. The concentration of a nutrient in a growth medium can determine (a) the amount of biomass formed in the culture, (b) the rate at which biomass is formed, or (c) both. Explain why nutrient concentration can determine either total growth or growth rate.

6. Plot a graph that illustrates the effect of temperature upon the growth rate of a microorganism.

7. Distinguish between psychrophiles, mesophiles, and thermophiles. Cite environments which are likely to favor the growth of each group.

8. What physiological properties of bacterial species prevent their growth (a) below a minimum temperature and (b) above a maximum temperature?

9. Plot the growth rate of (a) "normal", (b) osmotolerant, and (c) osmophilic organisms as a function of water activity.

10. What physiological mechanisms are used by osmotolerant and osmophilic microorganisms to cope with low water activity?

11. Contrast the oxygen requirements and oxygen sensitivity of (a) obligate aerobes, (b) facultative aerobes, (c) microaerophilic species, (d) aerotolerant anaerobes, and (e) obligate anaerobes.

12. What forms of elemental oxygen are potentially toxic to cells? What enzymes do microbes possess that detoxify these forms?

ANSWERS

Completion

1. logarithm of cell number; 2. turbidity measurement; 3. rate, yield; 4. exponential; 5. dilution rate; 6. protein synthesis; 7. RNA synthesis; 8. ATP; 9. nutrient depletion; 10. saturated; 11. 5 to 9; 12. decreases; 13. compatible solute; 14. superoxide dismutase; 15. obligate anaerobe.

Multiple choice

1. C; 2. B; 3. E; 4. A; 5. C; 6. A; 7. C; 8. A; 9. D; 10. B; 11. B; 12. B; 13. E; 14. D.

Matching

1. J; 2. C; 3. H; 4. K; 5. A; 6. F; 7. G; 8. B; 9. D; 10. I.

Discussion

1. See text Sections 5.1 and 5.2

2. See text Section 5.2

3. See text Section 5.4

4. See text Section 5.3

5. See text Section 5.5

6. See text Section 5.7

7. See text Section 5.7

8. See text Section 5.8

9. See text Section 5.10

10. See text Section 5.10

11. See text Section 5.11

12. See text Section 5.11

Chapter 6
MACROMOLECULES AND MOLECULAR GENETICS

OVERVIEW

Chapter 6 (pages 180-228) concerns itself with information flow from DNA to proteins -- how it occurs. Chapter 7 will look at how this flow of information is controlled. The central problem is how the information in a **gene** (a sequence of nucleotides) is converted into a sequence of amino acids -- a polypeptide chain. (Some regions of DNA do not code for proteins; rather they code for types of RNA.) A second problem covered in this chapter is the replication of the DNA genome, a copy of which must be passed to each cell during cell division.

CHAPTER NOTES

In eukaryotic and prokaryotic systems, DNA serves as a **template** from which information can be faithfully copied by way of complementary base pairing for both DNA replication and protein synthesis. **Three bases code for one amino acid**. There are 20 different amino acids that go

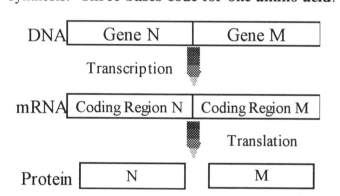

into the makeup of most proteins. However, over 100 different amino acids are known to exist. The information is coded in the base sequence of purine and pyrimidine bases on the polynucleotide chain. In nucleic acids, phosphate forms covalent bonds with the number 3 carbon of the ribose in one nucleotide, and the number 5 carbon of ribose in the adjacent nucleotide. This gives the molecule a polarity, we can identify a **3' end** and a **5' end**. The DNA molecule is composed of four different bases, **adenine**, **quanine**, **cytosine**, and **thymidine**. The important point to remember is that new RNA and DNA molecules are both synthesized in the 5' to 3' direction. The processes that underlie growth and division in a living cell include: **replication** where the DNA is duplicated; **transcription** where the information contained in the DNA is transferred to RNA; and **translation** where the information contained within the RNA is converted to amino acids in sequences that result in proteins. The one-way transfer of information from nucleic acids to proteins is referred to as the **central dogma of molecular biology**.

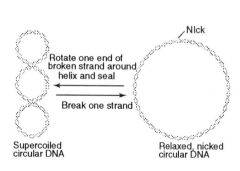

The bacterial chromosome, while simpler than eukaryotic systems, is a **double helix** that forms a closed circle. This structure has several important consequences for information flow. First, the twist of the double helix gives the molecule two grooves. Proteins that bind to DNA and control gene expression interact with the **major groove**. Second, the circular DNA is generally twisted in the cell -- that is, it is **supercoiled**. This twisting will make it difficult to gain access to single DNA strands to read them as templates. Therefore, enzymes called **topoisomerases** control the degree of twisting. Topoisomerase II causes the super coiling, whereas topoisomerase I will allow the relaxation of the coil. When a supercoiled molecule become nicked, it loses it supercoiling.

To this point, we have considered DNA structure on a gross level. However, the localized nucleotide sequence can give the molecule **secondary structure**, in which the nucleotides on 1 DNA strand can pair with each other. Such **stem and loop** structures may occur if the sequence contains **inverted repeats**. These structures are important in that they are recognition sites for proteins that bind to DNA. Long DNA molecules are normally quite flexible but stretches of DNA less than 100 bp are much more rigid. Certain sequences, however, result in bends in the DNA. Such **bent DNA** often involves several runs of five or six adenines in the same strand, each separated by four of five bases (see Figure 6.9).

The genetic material of eukaryotic cells has several important differences from that of prokaryotes. Of course, you already know that the genome in eukaryotes consists of a number of **chromosomes**. The major difference is that eukaryote chromosomes contain linear DNA molecules enclosed in a nuclear membrane. As a consequence, messenger RNA synthesis (**transcription**) and protein synthesis (**translation**) are physically separated in eukaryotes but are intimately coupled in prokaryotes. Thirdly, the genes in eukaryotic DNA have regions that do not code for protein (**introns**) among sequences that do (**exons**). During passage of **the primary transcript** or **pre-mRNA** from the eukaryotic nucleus, the molecule is processed; that is, the introns are excised. The excision of introns may not be catalyzed by a protein, but by small RNA molecules, called **ribozym**es, that internally cleaves and rejoins RNA molecules at specific, recognized sequences.

The box on pages 190-195 provides detailed information about some of the methods used to study DNA These methods include the purification of DNA -- cells are lysed, protein is denatured and removed, and RNA is destroyed enzymatically. DNA molecules can be separated by **density-gradient centrifugation** or by **gel electrophoresis**. The basis for density differences between DNA molecules is the number of guanine-cytosine base pairs (**GC content**). These have 3 hydrogen bonds per base pair versus 2 for adenine-thymine pairs. The extra hydrogen bond holds the two strands closer together, thereby making a denser structure. When centrifuged at high speed, the molecules "float" at the place in the tube where their density equals that of CsCl in solution. In electrophoresis, the negatively charged DNA molecules are attracted to the positive pole of an electrical field. Their rate of movement depends upon their size, as they must pass through pores in an **agarose gel**. DNA can be detected by **fluorescence** after the DNA is reacted with the dye **ethidium bromide**. DNA can be made **radioactive** using a number of different methods to incorporate ^{32}P into the DNA molecule. The structure of the DNA can be estimated by determining its melting point -- the melting point reflects the G + C content of the DNA. A more exacting description of the DNA structure can be determined by using DNA sequencing methods. The nucleotide sequence in DNA molecules can be compared by several methods. In **hybridization** techniques, the double helix is separated into single strands by heating. If one mixes different single stranded DNAs together, they will reform double strands in regions where the sequences are complementary. The hybridization is detected because one of the DNAs is made radioactive. Hybridization can be used to compare similarity of entire chromosomes, or can be restricted to a small piece of DNA, called a **probe**. Molecules can also be compared by directly determining the DNA sequences by chemical (Maxam-Gilbert) or enzymatic (Sanger) methods.

Two hydrogen bonds are formed by each AT pair, whereas three hydrogen bonds are formed by each GC pair. At room temperature and at physiological salt concentrations, DNA remains in a double stranded form. However, at increased temperature, the hydrogen bonds break and the strands separate. This process is known as **melting**. We already know that organisms vary in their relative content of G+C, relative to A+T. When a DNA molecule contains more GC pairs, the hydrogen bonding holding the strands together is stronger and a higher temperature is required to melt the DNA. The melting temperature, which is a characteristic of a given DNA molecule can be measured experimentally. **Hybridization** the process that occurs when complementary, single stranded DNA molecules are cooled slowly re-forming double stranded molecules.

The size of the chromosome in prokaryotes varies considerably, but is smaller that the chromosomal component in eukaryotic cells. In eukaryotic cells, the DNA contains a special sequenced called a **telomere** at each end a **centromer** in somewhere in between. The overall structure of a eukaryotic chromosome consists of a series of compacted, highly folded structures called **nucleosomes**. Each nucleosome contains about 200 base pairs. In contrast to prokaryotes, eukaryote DNA contains three classes of DNA. **Single copy DNA** encodes the main proteins of the cells whereas **moderately repetitive DNA** is found in a few to a large number of copies and encodes some major macromolecules in the cells (e.g., histones, immunoglobins, rRNA and tRNA). **Highly repetitive (satellite) DNA** is found in very high copy number, and comprises about 20-30% of the total DNA. The function of this class of DNA is not known.

A number of genetic elements are not part of the chromosome. These include **viruses**, **plasmids**, the genomes of **mitochondria and chloroplasts**, and **transposable elements**. Viruses are genetic elements -- either DNA or RNA, the control their own replication and transfer from cell to cell; viruses have been found for most prokaryotic and eukaryotic cells. Plasmids, which are often found in prokaryote cells but have been observed in eukaryotes, are small genetic elements that exist and replicate separately from the chromosome. The majority are double stranded and most are circular. Transposable elements are pieces of DNA having the ability to move from one site to another on a chromosome. They have been found in both prokaryotes and eukaryotes.

Restriction endonucleases are a class of highly specific enzymes that attack and break DNA at specific base sequences. These enzymes recognize unique **palindrome sequences** and make a double-stranded cut. The function of these enzymes in the cell is to destroy foreign DNA (such as from viruses) which is introduced into the cell. How do the restriction enzymes specifically recognize foreign DNA? A **modification enzyme** alters the nucleotides of the host DNA in the palindrome recognized by the restriction enzyme thereby blocking it from entering the enzyme's active site. A large number of restriction enzymes are now known and are available for use. These enzymes are thought to protect a cell from foreign DNA, such as a virus. Restriction enzymes have been important in molecular biology because they provide a convenient way to specifically cut large DNA molecules into smaller, manageable pieces.

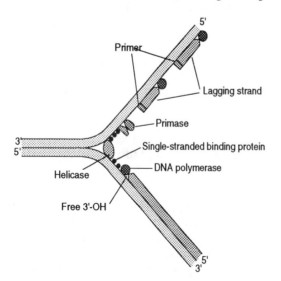

DNA replication involves breaking hydrogen bonds in the double stranded molecule to expose two **templates**, and choosing the correct base sequences in the newly synthesized strands by **complementary base pairing**. The enzyme, **DNA polymerase III** catalyzes the formation of covalent bonds between nucleotides in *E. coli*. Note however that it does not choose the specific nucleotide; that is determined by the template. An important characteristic of this enzyme is that it can only add nucleotides to a pre-existing nucleic acid chain. A small piece of RNA is first synthesized by primase, to act as a **primer** for DNA polymerase. This RNA is later degraded, and the gap filled in with DNA by **DNA polymerase I**.

The new strand of DNA is always synthesized in the 5' to 3' direction. This creates a problem, in that the template strand that runs in the 5' to 3' direction is oriented backwards for continuous synthesis of the new strand. In this case, replication proceeds **discontinuously**; that is, as a bit of the double helix is opened (the **replication fork**), DNA synthesis begins at the downstream end of the opened helix on one strand, and proceeds back to the spot where the previous fragment was initiated. The fragments are joined by **DNA ligase**. Note that topoisomerases are important in unwinding the DNA to allow the double helix to open for replication. The template strand that runs in the 3' to 5' direction is synthesized as large fragments and is known as the **leading strand**. The other, the **lagging strand**, is synthesized in small fragments (**Okazaki fragments**) and joing together by **DNA ligase**.

There are three types of RNA: (1) **ribosomal RNA**, structural components of ribosomes, (2) **transfer RNA**, an adapter molecule in protein synthesis, and (3) **messenger RNA**, which carries the coding information for a protein from the DNA gene to the ribosome, where the protein is synthesized.

RNA synthesis differs from DNA replication in that discrete starting points of genes must be recognized. These sites are called **promoters**, and **RNA polymerase** can bind to them. RNA polymerase consists of a number of subunits; the σ **subunit** is most important in recognizing promoters. Promoter sequences can vary, but two regions seem important in that they are conserved in most promoters. These are 10 and 35 bases upstream from where RNA synthesis begins. After polymerase binding, the DNA double helix unwinds, and RNA is synthesized using only one of the DNA strands as a template. Thus, the nucleotide sequence in the single-stranded RNA is determined by complementary base pairing. As in DNA synthesis, new nucleotides are added in the 5' to 3' direction. Termination of RNA synthesis is signaled by a **hairpin** structure in the RNA (secondary structure resulting from intrastrand base pairing).

In prokaryotes, genes involved in similar metabolic pathways are often physically linked. For example, all the genes coding for the enzymes of histidine biosynthesis are next to one another. When RNA polymerase binds to the promoter at the beginning of this region, it synthesizes one messenger RNA molecule that contains the coding information for all the proteins (**polygenic mRNA**). This provides a convenient way for the cell to coordinate synthesis of all these enzymes at once, by regulating binding to the one promoter.

The mRNA transcribed from a gene on DNA contains the information for specifying a protein. But how is it translated from the language of the four nucleotides to that of the twenty amino acids? Obviously, each nucleotide cannot represent a specific amino acid, because there are not enough different nucleotides. To specify all the amino acids, the nucleotide sequence must be read in groups of three (triplet **codons**). There are 64 unique codons, because there is a choice of four bases in each of the three positions. All but three specify an amino acid -- the others are 'stop' codons that terminate translation. Table 5.4 illustrates the correspondence between mRNA codons and the amino acids they specify (the **genetic code**). Notice that in many cases the third base in the codon is irrelevant -- the same amino acid is specified no matter what nucleotide is in this position.

The codons are read on ribosomes, which consist of two subunits, each of which are large complexes of RNA and proteins. The 30S subunit binds to the beginning of a mRNA, and the **start codon** (always AUG) is located. Note that for the proper protein to be produced, the mRNA must be translated in the proper **reading frame**. There is no internal 'punctuation' in the message to designate the beginning of codons. A shift of one base would produce a completely different polypeptide.

The correct amino acids are selected by using transfer RNAs as adapter molecules. The **anticodon loop** of a tRNA has a group of three nucleotides complementary in sequence to a specific codon. The particular amino acid specified by the complementary codon is covalently linked to one end of the tRNA. This linkage was catalyzed by one of twenty **aminoacyl-tRNA**

synthetases, in which the active site of the enzyme recognized the amino acid and only those tRNAs with anticodons complementary to codons for this amino acid. (Each synthetase recognizes the appropriate tRNAs from sequences in the acceptor stem of the tRNA molecule.) Therefore, the insertion of the correct amino acid during protein synthesis on ribosomes depends upon complementary base pairing between the mRNA codon and the tRNA anticodon carrying an amino acid.

The proteins and RNA in ribosomes are responsible for the recognition steps and peptide bond formation involved in protein synthesis. Both eukaryotic and prokaryotic cells synthesize proteins on ribosomes, and the mechanisms are generally similar. However, there are differences in these ribosomes, as indicated by the effect of certain antibiotics on prokaryotic, but not eukaryotic ribosomes.

SELF TESTS

COMPLETION

1. The synthesis of DNA and RNA always proceeds in the _____ direction.

2. Stem and loop secondary structures can be formed by DNA sequences that are _____.

3. Nucleic acid molecules can be separated on the basis of size by _____.

4. Sequence similarities between different DNA molecules can be determined by analyzing the sequences directly or by _____.

5. Restriction endonucleases make double stranded cuts in DNA at _____

6. The non-coding intervening sequences in eukaryotic genes are called _____.

7. Termination of mRNA synthesis can be signaled by a _____ structure in the RNA molecule.

8. Rifamycin inhibits _____ synthesis by binding to the beta subunit of _____.

9. There is a great deal of secondary structure in _____ RNA molecules.

10. Proteins that are secreted outside the cell membrane contain a _____ at their N-terminal end.

KEY WORDS AND PHRASES

-10 region (p 210)	-35 region (p 210)
actinomycin (p 212)	autoradiography (p 191)
buoyant density (p 190)	complementary base pairing (p 185)
D-loop of tRNA (p 216)	density gradient (p 190)
DNA gyrase (p 187)	DNA helicase (p 203, table 6.4)
ethidium bromide (p 191)	exons/introns (p 183)
formylmethionine (p 220)	genetic code degeneracy (p 222)
genetics (p 181)	kilobase (p 185)
lagging strand (p 203)	leading strand (p 203)
nonsense codons (p 222)	Okazaki fragment (p 04)

open reading frame (p 223)	operator (p 212)
operon (p 212)	origin of replication (p 203)
palindrome (p 199)	polyA tail (p 214)
polysome (p 220)	post-translational (p 221)
Pribnow box (p 210)	primary RNA transcript (p 213)
proofreading (p 205)	repetitive DNA sequences (p 198)
restriction enzyme mapping (p 201)	ribosome acceptor site (p 220)
ribosome peptide site (p 220)	rifamycin (p 212)
RNA cap (p 214)	satellite DNA (p 198)
secretory proteins (p 221)	semiconservative replication (p 201)
Shine-Dalgarno sequence (p 214)	signal sequence (p 221)
split genes(p 198)	start codon (p 222)
stem loop structure (p 188 and 210)	wobble in codons (p 222)

MULTIPLE CHOICE

1. DNA serves as a template for the synthesis of _____ by complementary pairing.

      (A) RNA and protein   (D) protein

      (B) DNA and RNA   (E) glycocalyx

      (C) DNA, RNA, and protein

2. During transcription

      (A) initiation occurs at a site recognized by σ factor

      (B) nucleotides are polymerized by DNA polymerase

      (C) only single gene-sized mRNA molecules are synthesized

      (D) both DNA strands of a single gene are used as templates

      (E) thymine in RNA pairs with adenine in DNA

3. Which of the following statements are INCORRECT?

      1. The genetic code is contained in 20 different codons.

      2. All codons specify an amino acid.

      3. Most codons pair with a specific anticodon of tRNA.

      4. Codons are made from the 4 base alphabet A,G,C,T.

      5. Different codons may code for the same amino acid.

      (A)1,4,5 (B) 2,4,5 (C) 1,2,4 (D) 1,3 (E) 2,3,4

4. Promoter regions are nucleotide sequences which

    (A) are involved in the initiation of transcription

    (B) are involved in transcription termination

    (C) contain the code for 1 mRNA molecule

    (D) are important to the translation process

    (E) specify amino acids

5. Which of the following events does NOT occur during translation?

    (A) Amino acids are joined through peptide bonds.

    (B) Codons are read in sequence to form a protein.

    (C) Amino acids are added to a tRNA after the tRNA binds to its specific codon on the ribosome.

    (D) The growing protein chain is transferred from a tRNA in the P site to a charged tRNA in the A site of the ribosome.

    (E) The newly synthesized protein is released when the ribosome reaches a stop codon.

6. Which of the following processes requires a primer:

    (A) protein secretion   (D) DNA synthesis

    (B) transcription   (E) all of the above

    (C) translation

7. DNA molecules can be separated on the basis of differences in size by:

    (A) agarose gel electrophoresis

    (B) restriction endonucleases

    (C) density gradient centrifugation

    (D) the Maxam-Gilbert procedure

    (E) the Sanger dideoxy procedure

8. The proofreading of newly synthesized DNA, to excise incorrect nucleotides which have been inserted, is done by:

    (A) a restriction endonuclease

    (B) DNA gyrase

    (C) DNA ligase

    (D) DNA polymerase III

    (E) Any of the above enzymes

9. The _____ strand of DNA is synthesized as small pieces of DNA called Okazaki fragments then joined together during DNA replication.

    (A) reverse strand

    (B) lagging strand

    (C) forward strand

    (D) reverse strand

10. A restriction endonuclease such as EcoRI functions by:

   (A) restricting the movement of ribosomes

   (B) combining with DNA at specific sequences of bases and cutting the strand

   (C) combining with RNA at specific locations and cutting the strand

   (D) controlling the spread of palindromes

11. Methylation of specific bases within a strand of DNA protects it against:

   (A) its own restriction endonuclease

   (B) non-specific protein attachment

   (C) other organisms restriction endonucleases

   (D) base pairing

   (E) more than one of the above

12. DNA polymerases do all of the following except:

   (A) synthesize new DNA in the 5' to 3' direction

   (B) catalyze the addition of nucleotides to free hydroxyl group

   (C) initiate a new strand of DNA

   (D) copy from an existing DNA template

   (E) work in a semiconservative process

13. Initiation, early replication and synthesis of DNA requires the following:

   (A) RNA primer region on the DNA

   (B) an origin of replication

   (C) the presence of helicases

   (D) DNA polymerase III

   (E) all of the above

14. DNA differs from RNA in that:

   (A) RNA is composed of the sugar ribose

   (B) RNA uses the base uracil

   (C) RNA tends to be single stranded

   (D) all of the above

   (E) none of the above

15. In the formation of proteins the ribosomes play a key role. The overall steps in the protein synthesis process are:

    (A) termination-release and polypeptide folding, elongation

    (B) elongation, initiation, termination-release and polypeptide folding

    (C) polypeptide unfolding, elongation, initiation

    (D) initiation, elongation, termination-release and polypeptide folding

    (E) initiation and polypeptide folding

MATCHING

I. Match the enzyme with its action

Enzyme	**Action**
1. DNA methylase	(A) degrades DNA to nucleotides
2. DNAse	(B) links DNA molecules
3. DNA helicase	(C) adds methyl group to DNA
4. DNA gyrase	(D) attaches nucleotide, proofreads each insertion
5. DNA polymerase III	(E) cuts DNA at a specific sequence
6. primerase	(F) binds to DNA near the replication fork
7. DNA polymerase I	(G) increases twisting of DNA
8. DNA ligase	(H) attaches nucleotide, removes RNA primer
9. restriction endonuclease	(I) makes short RNA using a DNA template

DISCUSSION

1. Discuss the methods that are available to separate DNA molecules. What physical differences between molecules are the basis for the separation techniques? What methods are used to detect separated DNA molecules, with the goal of detecting (a) all molecules or (b) specific sequences?

2. List 3 enzymes that are important in DNA replication. What function does each serve in this process?

3. Describe 3 cases where the formation of secondary structure in a nucleic acid has a biological function.

4. Describe the signals for both initiation and termination of (a) transcription and (b) translation.

5. For a DNA sequence (3')TACCTTAGGCTACGTAGA(5') : (a) Write the mRNA sequence synthesized from it. (b) What tRNA anticodons would pair with the mRNA. (c) What amino acid sequence would be specified by this sequence.

6. Compare gene expression in eukaryotic and prokaryotic cells.

7. What is an operon?

8. Ribosomes are composed of RNA and proteins. In which steps in protein synthesis does the ribosomal RNA play an active, functional role? What steps in protein synthesis can be inhibited by antibiotics?

9. Which amino acids have the fewest number of codons in the genetic code? Which have the greatest number? In what proportion of cases are the presence of multiple codons due to differences only in the third position?

10. Detail the roles that RNA (m, t, and rRNA) play in cell function.

ANSWERS

Completion

1. 5' to 3'; 2. inverted repeats or palindromes; 3. electrophoresis; 4. hybridization; 5. palindromes; 6. introns; 7. stem and loop; 8. mRNA, RNA polymerase; 9. transfer RNA; 10. signal sequence;

Multiple choice

1. B; 2. A; 3. C; 4. A; 5. C; 6. D; 7. A; 8. D; 9. B; 10.B; 11.A; 12.C; 13. E; 14. D; 15.D.

Matching

1. C; 2. A; 3. F; 4. G; 5. D; 6. I; 7. H; 8. B; 9. E.

Discussion

1. See box on page 190 of text.

2. See Table 6.4.

3. See text Sections 6.7 - 6.9

4. See text Sections 6.6 and 5.9.

5. (a) AUGGAAUCCGAUGCAUCU

   (b) UAC,CUU,AGG,CUA,CGU,AGA

   (c) methionine, glutamate, serine, aspartate, alanine, serine.

6. See text Section 6.6 and 6.7.

7. See text Section 6.6.

8. See text Section 6.9.

9. See Table 6.6.

10. See text Sections 6.7 - 6.9.

REGULATION OF GENE EXPRESSION

OVERVIEW

Chapter 7 (pages 228-249) considers **gene regulation**. Although the genome of a cell contains the coding capacity for many proteins, not all are **expressed** at one time. Moreover, even when an enzyme is present in the cell, it may not be active under a given set of conditions. This chapter investigates the mechanism of these regulatory processes.

CHAPTER NOTES

Although many enzymatic reactions occur during a single cycle of cell, not all occur the same extent. Some compounds are need in large amounts, whereas others are required, but in much lessor quantities. Moreover, some proteins will be needed in greater quantities under some conditions and lessor quantities, while others will be needed in roughly constant amounts under most growth conditions. Because cells will need to make maximum use of their available resources, these processes need to be regulated.

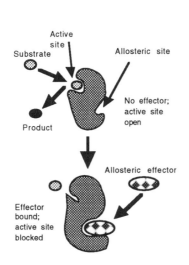

Biosynthetic pathways may be controlled by **feedback inhibition**, in which the **end product** of the entire pathway inhibits the activity of the **first enzyme** *unique* to the pathway not the gene. If the first reaction does not occur, the subsequent enzymes are "starved" of their substrates, and the end product is not synthesized. The end product binds to the first enzyme at a site different from the **active site**. This second binding site on the protein is called the **allosteric site**. Binding of the **allosteric effector** changes the three dimensional structure of the protein, and especially the conformation of the active site. Thus, substrates no longer bind to the active site, and the enzymatic reaction does not proceed until the **feedback inhibitor** leaves the allosteric site.

A second mechanism to alter enzymatic activity is by **covalent modification**. A modifying enzyme catalyzes the addition (via a covalent bond) of a phosphate group, methyl group, or nucleotide to a biosynthetic enzyme. This alters the conformation of the active site of the biosynthetic enzyme, and thereby alters its activity. The modifying group can be removed by another enzyme, to restore the original state of activity.

Regulation of enzyme amount occurs at the gene level by a process termed **induction** or **repression**. Because they are so energetically costly to make, regulating the amount and types of proteins synthesized is very effective in optimizing growth. Control occurs at the level of transcription. Simply, the cell regulates the binding of RNA polymerase to promoters coding for specific proteins. There should be some sort of signal that indicates whether the protein is needed under the current environmental conditions. Usually this is a small molecule (an **effector**). However, the effector does not bind to the promoter itself. Rather, there are a series of **regulatory proteins** that can bind to DNA at specific sites (**operators**) that are located near promoters. The regulatory proteins contain **allosteric sites** to which the effectors can bind. The change in protein conformation caused by effector binding may either increase or decrease the affinity of the regulatory protein for an operator. In those cases where the affinity is increased by binding of effector to regulatory protein so that the complex prevents RNA polymerase from transcribing the gene, the effect is called **repression**. If effector binding alters the **repressor protein** so that it no longer prevents RNA polymerase binding to the promoter, the effector is responsible for **enzyme induction**. There is a lot of terminology associated with this subject, but the important point to keep in mind is the mechanism -- a regulatory protein can physically make a promoter inaccessible to RNA polymerase by binding to a nearby operator. The role of effector molecules is to activate or deactivate the capacity of regulatory proteins for operator binding.

In the mechanisms described above, the regulatory protein exerts a **negative control** on gene expression -- its binding prevents mRNA synthesis. For this reason, it can also be called a repressor protein. There are also cases where the binding of a regulator protein near the promoter enhances RNA polymerase binding -- this is called **positive control**.

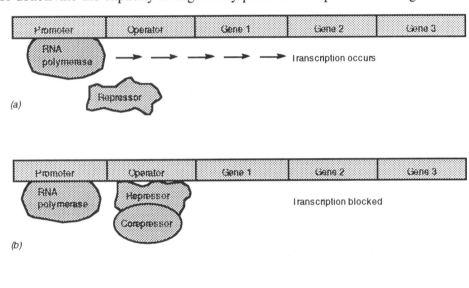

Attenuation is a control mechanism found only in prokaryotes because it requires close coupling between transcription and translation. It operates in operons for amino acid biosynthetic enzymes. The transcription of the structural genes is not begun unless the cells have a low level of the amino acid in question. This condition is signaled as ribosomes translate the **leader mRNA**; there are codons for the specific amino acid in the message. If there is little charged tRNA for this amino acid, the ribosomes stall here, and physically prevent the transcription termination signal (see Figure 7.17) from forming at the end of the leader sequence, so that transcription continues into the structural genes.

A cell can regulate many different genes at one time. This higher mode of cellular operation is

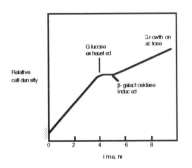

termed **global regulation**. This term describes the control mechanisms the cell exerts over many different genes in response to changes in its environment. **Diauxic growth** on glucose and lactose, a type of catabolite repression, is a consequence of global regulation. When the cell is presented with the two nutrient sources, it will suppress the formation of enzymes for lactose and use glucose. As the glucose is depleted, the repression is lifted and enzyme production for the second nutrient is started. The process reflects the binding of a complex between **catabolite activator protein (CAP)** and

cyclic AMP to the promoter for β-galactosidase, the enzyme formed to use lactose. In order for the process to occur, cyclic AMP must bind to CAP. In cells growing on glucose the level of cyclic AMP is low and CAP is not activated. Other global regulation systems include control over aerobic and anaerobic respiration, nitrogen utilization, heat shock and oxidative stress.

In the nature, bacteria are exposed to wide fluctuations in environmental conditions, such as temperature, pH, oxygen, nutrient availability, and inorganic ions. Response to environmental conditions is often brought about by **two component systems**. Two component systems consist of **sensor proteins**, which is often a **kinases (sensor kinase)**, and **response regulator proteins**. The kinase phosphorylates itself (**autophosphorylation**) in response to environmental conditions, then transfers the phosphate to the response regulator protein. The response regulator protein is typically a **DNA binding protein** that acts as a **repressor** in the phosphorylated form while being inactive in the unphosphorylated form. A "third component" of two component systems often includes a **phosphatase** that removes the phosphate from the response regulator to stop the response. There are many well documented examples of two component systems, including nitrogen assimilation in *E. coli*, nitrogen fixation in *Klebsiella* and sporulation in *Bacillus*. *E. coli* is known to contain at least 50 two component systems, and closely related systems have been found in lower eukaryotes, such as *Saccharomyces cerevisiae*.

Chemotaxis involves a number of proteins that sense changes in **attractant** or **repellent** concentration over time. **Sensory proteins** are **methyl-accepting chemotaxis proteins (MCP's)**. Each **MCP** can sense a variety of compounds. MCP's bind attractants or repellents directly or indirectly through interactions with **periplasmic binding proteins**. Bind of the attractant or repellent causes a series of interactions with cytoplasmic proteins that results in *reversal* of **flagellar rotation**. In most cases, counterclockwise rotation results in cell movement (**a *run***) whereas clockwise movement results in a **tumble**. Positive changes in attractant concentration (or negative changes in repellent concentration) extend the duration of the run, whereas the reverse shortens the duration of the run and induces a tumble.

SELF TESTS

COMPLETION

1) *E. coli,* when grown in a medium that contains both lactose and glucose, will use the _____ first.

2) The two major mechanisms of regulation enzyme activity in a cell include control preexisting enzyme _____ and regulation of the _____ of the enzyme synthesized.

3. _____ are enzymes that catalyze the same reaction.

4. A set of genes with one control unit for expression is an _____.

5. The enzyme β-galactosidase is synthesized in *E. coli*, only when lactose is present in the medium. This is an example of _____.

6. A substance that represses enzyme production is called a _____.

7. The protein that brings about repression is called a _____ protein.

8. In *E. coli*, the enzymes for utilizing lactose are controlled by a repressor protein, while, in contrast, a _____ protein is required for induction of the enzymes required for maltose utilization.

9. During attenuation of the tryptophan operon, the key feature of the peptide encoded by the leader sequence is a pair of _____ residues near the terminus of the peptide.

10. During catabolic repression, the synthesis of enzymes required for use of a number of sugars is repressed when _____ is present in the medium.

11. In two component systems, the first component, the sensor protein, is often a _____.

KEY WORDS AND PHRASES

β-galactosidase (p 235)	activator protein (p 238)
active site (p 231)	allosteric site (p 231)
allostery (p 231)	alternative sigma factors (p 243)
attenuation (p 239)	attenuator (p 239)
catabolite activator protein (p 242)	catabolite repression (p 241)
chemotaxis (p 245)	corepressor (p 236)
covalent modification (p 232)	cyclic AMP (p 242)
diauxic growth (p 241)	DNA binding protein (p 232)
effector (p 231 and p 236)	enzyme induction (p 235)
enzyme repression (p 235)	feedback inhibition (p 230)
global control systems (p 241)	glucose effect (p 241)
heat shock proteins (p 243)	histones (p 233)
inducer (p 236)	isozyme (p 231)
kinase or sensor kinase (p 244)	leader sequence (p 239)
leucine zipper (p 234)	maltose regulon (p 237)
chemotaxis proteins (p 245)	negative control (p 237)
operator region (p 237)	operon (p 237)
phosphatase (p 244)	positive control (p 237)
product inhibition (p 230)	regulatory protein (p 232
regulon (p 238)	repressor protein (p 236)
response regulator protein (p 244)	runs and tumbles (p 246)
sensor protein (p 244)	signal transduction (p 244)
substrate (p 230)	termination of transcription (p 239)
transcription pause site (p 240)	two component systems (p244)
zinc finger (p 234)	

MULTIPLE CHOICE

1. Biochemical pathways may be regulated by

    (A) controlling the level of enzyme synthesis.

    (B) controlling the level of enzyme activity.

    (C) Both A and B

2. The mechanism of control whereby the end product of the pathway inhibits the first enzyme in the pathway is called

    (A) product inhibition   (B) feedback inhibition

    (C) enzyme repression   (D) enzyme induction

    (D) feedback loop repression

3. The substance that initiates enzyme induction is called a (an) _____.

    (A) inductor   (B) co-effector

    (C) initiator   (D) inducer

    (d) allosteric effector

4. In order for *E. coli* to utilize the sugar lactose, which of the following must be true?

    (A) Glucose must be absent or used up in the medium.

    (B) The level of cyclic AMP must be elevated in the cell.

    (C) β-galactosidase must be synthesized

    (D) The inducer must be bound the lac repressor protein.

    (E) All of the above must be true.

5. A major difference between the utilization of lactose and utilization of maltose in *E. coli* is that

    (A) maltose is required for biosynthetic activities, but lactose is not.

    (B) regulation of the maltose operon is an example of positive control while lactose regulation is an example of negative control.

    (C) maltose regulation does not require the presence of maltose in the medium.

    (D) the maltose utilization is not subject to catabolic repression.

    (E) All of the above are true.

6. A regulon

    (A) is a group of coordinately controlled operons.

    (B) is a synonym for an operon.

    (C) is a synonym for allosteric enzyme control

    (D) refers to a type of enzyme.

    (E) All of the above are true

7. The process of attenuation

    (A) is more common in biosynthetic pathways.

    (B) refers to fine control of enzyme synthesis.

    (C) can be accomplished by early termination of transcription.

    (D) in the tryptophan operon, is a function of the leader sequence.

    (E) All of the above are true.

8. The glucose effect is associated with

    (A) catabolic repression.   (B) the synthesis of tryptophan.

    (C) DNA synthesis.   (D) attenuation.

    (E) histidine biosynthesis.

9. When *E. coli* is grown in a medium containing several sugars, which of these would you expect to be used first?

    (A) lactose   (B) lactose

    (C) glucose   (D) sucrose

    (E) all should be used at the same time.

10. In a two component regulatory system, which of the following must be membrane-bound?

    (A) the sensor protein   (B) the response regulator protein

    (C) the repressor protein   (D) the phosphatase

    (E) All must be membrane bound

11. Which of the following are true of chemotaxis.

    (A) Attractants and/or repellents might be involved.

    (B) The cycle consists of alternate runs and tumbles, but when the movement is toward an attractant (or away from the repellent), the runs are longer.

    (C) The mechanism involves reversing the direction of flagellar rotation.

    (D) The methyl-accepting chemotaxis proteins are trans-membrane proteins.

    (E) All of the above are true.

MATCHING

Match process with its correct regulatory component

Process	Component
1. Feedback inhibition	A. cyclic AMP and the CAP protein
2. Catabolite repression	B. an allosteric enzyme
3. Enzyme induction by a repressor protein	C. a two component system
4. Enzyme induction by an activator protein	D. the leader sequence
5. Attenuation	E. the inducer
6. Signal transduction	F. MCP's
7. chemotaxis	G. the co-repressor

DISCUSSION

1. What are the advantages and disadvantages to a cell as a consequence of regulating a pathway by feedback inhibition as compared to enzyme induction.

2. Compare and contrast product inhibition and feedback inhibition.

3. Describe the differences between positive control and induction and repression as the terms related to regulation of transcription.

4. Based on what you know about energy generation in E. coli (Chapter 4), why would glucose be used preferentially to maltose or lactose?

5. Compare and contrast the terms regulon and global control system. Why is catabolic repression considered a global regulation system rather than a regulon?

6. Figures 7.16 and 7.17 show that tryptophan synthesis is regulated by both repression of transcription and attenuation. Explain this statement.

7. One would not expect to see attenuation by transcription termination, as in the *E. coli* tryptophan operon in a yeast. Why?

8. In *Bacillus subtilis*, one finds new and different sigma factors associated with the RNA polymerase as the cells begin to form endospores. Why?

9. Describe the process of chemotaxis in bacteria.

ANSWERS

Completion

1. lactose; 2. activity, amount; 3. Isozymes; 4. operon; 5. enzyme induction; 6. corepressor; 7. repressor; 8. activator; 9. tryptophan; 10. glucose; 11. kinase

Multiple choice

1. C; 2. B; 3. D; 4. E; 5. E; 6. A; 7. E; 8. E; 9. C; 10. A; 11. E:

Matching

1. B; 2. A; 3. G; 4. E; 5. D; 6. C; 7. F.

Completion

1. See text sections 7.1, 7.3, and 7.4.

2. See text section 7.1

3. See text section 7.3 and 7.4.

4. See text section 7.3 and Chapter 4.

5. See text sections 7.4 and 7.5.

6. See text section 7.5

7. See text sections 7.5 and 7.8.

8. See text section 7.8

9. See text section 7.7

Chapter 8
VIRUSES

OVERVIEW

Chapter 8 (pages 250-305) is concerned with viruses. Viruses are **non cellular genetic elements** that contain nucleic acid that is surrounded by a shell of protein. They are smaller than cells, ranging in size from 0.02 to 0.03 μm. In the extracellular state the virus particles, also know as **virions**, are metabolically inert. Because they have no metabolic machinery, these genetic elements must **infect** cells and use the host's ribosomes, enzymes, and catabolic potential to reproduce themselves. Bacterial, animal, and plant cells are susceptible to infection by specific viruses.

CHAPTER NOTES

Viral genomes are much smaller than bacterial systems: one of the largest viral genomes is 190 kilobases compared to 590 for the smallest bacterial system (see table 6.1). Viruses are not restricted to double-stranded DNA as their genetic material. The DNA may be single-stranded, or RNA as a single- or double-stranded molecule may serve as the genetic material. The viral genetic material is located within the particle and is surrounded with protein. This structure is called a **nucleocapsid**. These unusual forms of genetic material create unique problems in genome replication and production of mRNA. Genomes may also be **segmented** into several molecules within the protein coat.

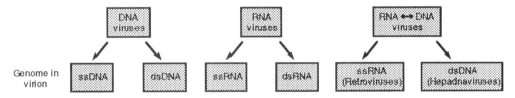

The protein coat consists of a number of protein subunits (**capsomeres**) which form either a helix or a twenty-sided shell (**icosahedron**) around the nucleic acid. These arrangements are geometrically the most efficient in minimizing the number of capsomeres necessary to enclose the genome, and in allowing similar protein-protein interactions to occur between identical protein subunits so that they may **self-assemble**.

These are the minimum requirements for a virion. Some viruses may have other structures such as tails or spikes, which are important in cell infection. Others, notably some animal viruses, have membranous **envelopes** surrounding the **nucleocapsid**. These are added as particles bud through the cell membrane. Thus, the lipid in the envelope is cell-derived, but the proteins may be virus-encoded ones inserted into the membrane during virus replication.

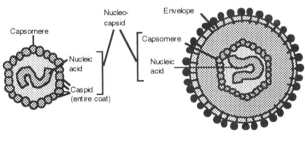

Viruses are obligate intracellular parasites. Therefore, to maintain them in the laboratory, it is necessary to grow host cells, which are then infected with virus. For bacterial viruses, this involves the normal culturing techniques you have learned. Animal cells are grown in **tissue culture**, on the surface of a dish flooded with a complex and highly nutritious medium.

Viruses are enumerated by spreading a suspension on the surface of a lawn of actively growing cells susceptible to the virus. As a virus particle infects, reproduces, and its progeny reinfect surrounding cells, a zone of clearing in the cell layer is produced. This **plaque** presumably originated from one virus particle, and therefore the number of virions in the original suspension is proportional to the number of plaques obtained.

Viruses have two types of reproductive cycles -- (a) **virulent** or **lytic** infection, in which new virus particles are made and released (burst) from the **host cell**. (b) The virus genome may become **integrated** into that of the host cell, and be replicated as part of the host as it grows. This process is called **lysogeny** in the case of bacteriophage, and **transformation** in animal viruses.

In either lytic or lysogenic process, the virus must enter the cell. This involves two steps. (1) *Attachment* to the surface of the cell. The type of cells a virus can infect is highly specific. A virus protein binds specifically to a cell receptor, which generally has some non-viral cell function. In some eukaryotic cells, virions are nonspecifically carried in by endocytosis. This also accomplishes the second stage of infection -- (2) *penetration*. In bacteriophage, the cell wall and membrane are breached by enzymatic degradation and injection of the genetic material into the cell. Note that the protein coat does not enter the cell. Its function is protection, and the nucleic acid is capable of directing virus replication.. The **eclipse period** -- where nucleic acid is separated from the protein coat -- follows infection. The host can protect itself in a number of ways. The first is to have no receptor sites for the virus to bind with. The second, found in prokaryotes, is the production of restriction enzymes, which attack the foreign nucleic acid.

In lytic infections the **latent period** when no virons are evident follows the eclipse period. The subsequent important events are (3) the take-over of host biosynthetic machinery to prevent synthesis of cellular proteins and enable expression of viral genes. (4) Replication of nucleic acid. (5) Synthesis of coat proteins. (6) Assembly of genomes into protein coats to produce ineffective particles. (7) Release from the cell. These processes make up a one-step growth response and occur rapidly after infection. The time-frame for the process can be 20-30 min in bacteria and 8-40 hours in animal cells.

In a **lysogenic infection**, the virus must (1) repress expression of the genes used to initiate lytic infections and (2) insert a double-stranded DNA copy into the host chromosome. The phage genes are held in the lysogenic stage until a switch or signal is detected at which time they convert to the lytic growth cycle described above.

The text describes a number of bacterial and animal viruses. Let us first summarize some general patterns among these before considering any unique aspects of particular viruses.

(1) Virus size. The smallest viruses have so little capacity for coding proteins in their genomes that they are highly dependent upon existing host enzymes for replication. Examples are SV40 and φ-X174. In the latter, economy is achieved with **overlapping genes**, in which some nucleotide sequences are read in more than one reading frame. As their genome size increases, virus structure and life cycle become more complicated.

(2) Termination of host biosynthetic activity. This may involve proteins that inhibit host RNA polymerase (T4 and T7 phage), synthesis of new RNA polymerases or sigma-like factors that do not bind to host promoters (T7 phage), inactivation of translation factors (poliovirus) or by degradation of host DNA (T4 phage). In T4, this means phage DNA must be recognized as different. It contains a unique nucleotide, **hydroxymethylcytosine**; of course, the phage genome must code for the enzymes unique to its synthesis.

(3) Transcriptional programs. Different sets of virus genes are transcribed at different periods of the infection cycle. Usually, the host RNA polymerase will transcribe the "early" genes, one of

which may code for a new virus-specific RNA polymerase or new sigma factors. These proteins shift transcription to "middle" or "late" genes necessary for genome replication, synthesis of structural proteins, and assembly. Lytic proteins are synthesized very late in the infection cycle, and lead to release of virus particles from the cell

(4) DNA replication. Replication mechanisms such as the **rolling circle** (lambda phage) or base-pairing of direct repeats at the ends of the genome (T4 and T7) can generate **concatemers** that contain many copies of the genome linked end-to-end in 1 large molecule. These are enzymatically cut into lengths that fit into the phage head. The length will be slightly greater than a complete genome. Thus, some sequences will be duplicated, and different genes may be at the molecule's ends in individual phages.

(5) RNA viruses. These have a unique problem, in that cells do not have enzymes that make RNA copies of an RNA molecule. Therefore, such an enzyme must either be contained in the virion (rhabdoviruses and reoviruses) or the RNA must immediately be translated to synthesize this enzyme (MS2 and poliovirus).

RNA bacteriophage (example: MS2) have RNA genomes, are very small, and have icosahedral structures with 180 copies of protein coat per virus particle and only infect bacteria that are gene donors. The RNA acts as mRNA and encodes only a few proteins. The virus uses the host's proteins to construct a number of viral materials. **Single-stranded icosahedral DNA bateriophages** (example: φ-X174) are small viruses with genomes composed of small pieces of single-stranded DNA in a circular configuration. **Single-stranded filamentous DNA bacteriophage** (example: M13) are linear but possess circular single stranded DNA. They are used extensively in genetic engineering work as they can be released from the cell without killing the cell. **Double-stranded DNA bacteriophages** (example: T7 & T3) have icosahedral heads, small tails, five different proteins in the head and three to six proteins in the tail. **Large double-stranded DNA bacteriophages** (examples: T2, T4) are structurally among the most complicated viruses. In addition to the **head** that contains the DNA, they contain a tail, through which the DNA is injected into the bacterium, and in the case of T4, tail fibers that are important for attachment to host cells.

Lambda phage is an example of a virus that has alternative replication cycles. It can infect a cell and undergo the lytic cycle, which ends with release of 100 or more new virons from a lysed cell. Alternatively, its DNA can **integrate** into the bacterial chromosome, where it is maintained as a **prophage** in a **lysogenic** state and replicated by the host as part of its DNA, until it is **induced** to excise itself and initiate the lytic cycle. The choice between these alternatives is made after injection of the DNA into the host cell. The host RNA polymerase binds to two promoters -- one for the lambda repressor gene, and the other for the *cro* gene. If lambda repressor protein is made more rapidly than Cro protein, the lytic cycle is repressed, because the repressor protein binds near promoters of lytic genes and prevents their expression. Cro protein represses expression of lambda repressor protein, and thereby permits expression of lytic functions. When lambda is integrated as a prophage in the chromosome, only the lambda repressor gene is expressed. This not only prevents the lytic cycle of this phage, but confers **immunity** on the host cell to infection by other lambda phage. If another lambda genome is injected, lambda repressor can bind to it to prevent expression of lytic genes.

The lytic cycle of lambda prophage is induced by inactivation of lambda repressor. The prophage is excised from the chromosome, and lytic genes are expressed.

Mu phage can also insert into the host cell chromosome. However, in contrast to lambda, which inserts at one specific site, mu has properties of a transposable element, and can insert at many sites in the DNA.

When animal cells are infected by viruses, the entire virus usually enters the cell by **endocytosis**. Although the consequence of virus production may be cell lysis, there may be continuous virus production without lysis. This is especially true of enveloped viruses, in which the lipid envelope is added to the nucleocapsid as the virus **buds** through the cell membrane.

Poliovirus is an RNA virus in which the genome is translated by ribosomes into 1 large polypeptide. This molecule is then **post-translationally cleaved** by a **protease** into the smaller, functional protein molecules. Poliovirus RNA molecules can function as "genomes", destined for packaging into protein coats, or as mRNA. The binding of VPg protein to the RNA determines which function a specific molecule will perform. In the single-stranded RNA rhabdoviruses, a similar distinction is made on the basis of size and sequence. The genome is not translated; rather, a series of shorter mRNA molecules are made using it as a template. Full-length copies of the genome are used as templates to synthesize new copies of the genomic RNA.

Rhabdoviruses are also **enveloped**. Some viral genes code for unique glycoproteins that are inserted into the lipid matrix of the cell membrane. Nucleocapsids align with these and bud through the membranes at these points, so that the lipid envelope contains virus-specific proteins.

Influenza virus is an example of an RNA virus with a **segmented** genome. When the virus reaches the nucleus, each molecule serves as a template for production of a unique mRNA molecule that codes for one virus protein. The segmented genome is analogous to the separate chromosomes of eukaryotic cells. In eukaryotes, new combinations of chromosomes can arise during meiosis. If more than one influenza virus infects a cell, the mixing of segments in the viral progeny can lead to an **antigenic shift**, in which the surface proteins of the new virus are altered.

Reoviruses are similar to influenza in that they have segmented RNA genomes, in which each segment specifies one protein, but differ in that their RNA is double-stranded. Infection is begun by binding to a cell surface protein whose normal function is as a **hormone receptor**.

SV40 is a small papovavirus that can carry out lytic infections, or integrate into host DNA (as in lysogenic bacteriophage). This latter process is called **transformation**, because the host cells exhibit very different properties in tissue culture. They behave like cancer cells, and indeed SV40 does cause cancer in some animals. In contrast to lambda phage, where infection of an individual cell may potentially lead to either the lytic or lysogenic cycle, the mode of SV40 infection is determined by the type of animal cell that is infected.

Herpesviruses cause a variety of human diseases. Especially noteworthy is their capacity for **latent infections**, in which symptoms only arise under stress. A unique morphological property is the **tegument**, a fibrous structure outside the nucleocapsid. This virus has a transcriptional program that controls the timing of gene expression. A unique characteristic of assembly of this enveloped virus is that the nucleocapsid is assembled in the nucleus, and the envelope is acquired by budding through the nuclear membrane, rather than the cytoplasmic membrane.

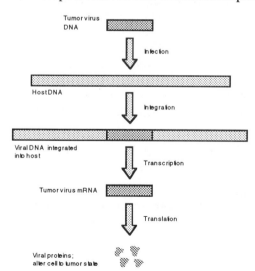

Pox viruses differ from other DNA viruses of animal cells in that they replicate in the cytoplasm rather than the nucleus. These extremely large viruses are medically important -- **smallpox** is a human pathogen that has been completely eliminated.

Retroviruses are of special interest because (1) they may cause diseases such as cancer or **AIDS** in animals or humans, and (2) they carry out a unique reaction. They can transform cells by integrating a copy of their genome into a host chromosome. What is unusual is that their genome is single-stranded RNA. Therefore, they must synthesize a double-stranded DNA molecule from this RNA to integrate into the host DNA. This reaction is catalyzed by the enzyme **reverse transcriptase**, a complex protein that contains 3 different enzymatic activities. The first uses RNA as a template to synthesize a complementary single-stranded DNA. The enzyme then hydrolyzes the RNA primer molecule (using a second enzymatic activity -- ribonuclease H activity), and finally synthesizes the complementary DNA sequence to form a double stranded molecule that can be integrated into the host genome.

Since reverse transcriptase is not a normal host enzyme, it is part of the virion that infects a cell. However, the reaction does depend upon a cellular molecule -- a tRNA that acts as a primer for reverse transcription. The DNA that is synthesized has **terminal repeat** sequences; its integration is analogous to that of bacterial transposons and the bacteriophage mu in that its insertion is not site-specific, and is catalyzed by an enzyme (in this case, the retrovirus **integrase**).

The tumor-causing potential of retroviruses is due to **oncogenes** that bring about cell transformation. DNA hybridization studies have shown that normal cells contain sequences similar to oncogenes. This suggests that these types of genes are fundamentally important in regulating cell growth, and that oncogenes have altered regulation that causes abnormal growth.

The mechanism whereby retroviruses introduce copies of their genetic information into animal cell genomes may prove useful to incorporate new genes into animal cells. An application would be to replace a defective gene that is responsible for a disease condition with a good copy of that gene. The retrovirus would serve as a **vector** for the **gene therapy**; the retrovirus would be genetically modified so that it would be incapable of replication itself.

SELF TESTS

COMPLETION

1. The _____ that make up the virus capsid associate by _____. This is possible because viruses have either _____ or _____ symmetry.

2. The uncontrolled growth of cells in an animal is called _____.

3. A _____, originating from a single virus particle is analogous to a bacterial colony.

4. During a lytic cycle of virus replication, the infectivity of the virus is lowest during the _____ period.

5. Animal viruses that do not interact with specific cell receptors infect cells by the cellular process of _____.

6. The viral genes that are expressed early in infection are generally transcribed by the _____ RNA polymerase.

7. A genetic alteration in the type of cells a virus can infect is a _____ mutation.

8. The phenomenon of _____ occurs when 2 viruses infect a cell and the genome of one is packaged into the capsid of the other.

9. If the infecting RNA of a virus can be translated directly, it is of the _____ sense.

10. In single-stranded DNA bacteriophages, mRNA is synthesized from a _____-stranded DNA molecule.

11. Bacteriophage coat proteins are generally made _____ in the infection cycle.

12. Nucleases produced by T4 phage do not degrade viral DNA because it is modified with _____ and _____.

13. Lytic growth of lambda prophage can be induced by destroying the _____.

14. Mutations in the bacterial genome occur during the life cycle of _____ phage.

15. Animal viruses are more likely to have _____, derived from _____ outside the nucleocapsid than are bacterial viruses.

16. In _____, the genome is a single-stranded RNA molecule that is directly translated by ribosomes.

17. It is essential that viruses with _____-sense _____ as their genetic material carry their own RNA polymerase into the cell they infect.

18. Animal cells can be transformed by viruses of these groups: _____ and _____.

19. _____ are unique DNA viruses of animals in that they replicate in the cytoplasm.

20. The capacity for animal cell transformation is a result of viruses containing _____.

KEY WORDS AND PHRASES

adenovirus (p 296)	bacteriophage (p 251)
benign (p 285)	burst size (p 260)
capsid (p 254)	capsomere (p 254)
early genes (p 274)	eclipse phase (p 260)
efficiency of plating (p 258)	envelope (p 254)
Epstein-Barr virus (p 294)	focus of infection (p 258)
hemagglutinin (p 289)	herpesvirus (p 294)
infection (p 251)	insertion element (p 280)
late genes (p 274)	latent infections (p 284)
latent phase (p 260)	lysogenic (p 275)
malignant (p 285)	metastasis (p 285)
neoplasm (p 285)	neuraminidase (p 289)
nonpermissive cells (p 293)	nucleocapsid (p 254)
one-step growth curve (p 260)	papovavirus (p 292)
permissive cells (p 293)	persistent infections (p 284)
phage (p 252)	picornavirus (p 286)
poxvirus (p 295)	prions (p 302)
prophage (p 275)	protein kinase (p 301)
reverse transcriptase (p 298)	RNA replicase of MS2 (p 267)
rolling circle replication (p 268)	temperate (p 275)
transposon (p 280)	virion (p 251
viroids (p 302)	virulent (p 275)
virus symmetry (p 254)	

MATCHING

Match the term with its definition.

1. icosahedron	(A) lysogeny effect
2. enveloped virus	(B) proteins that aid in self-assembly
3. molecular chaperones	(C) nucleocapsid enclosed in a membrane
4. neuraminadases	(D) kills host
5. virus infection unit	(E) x-rays, ultraviolet radiation
6. eclipse phase	(F) virus sheds protein coat
7. virulent virus	(G) virus enzyme that attacks glyco-proteins
8. temperate virus	(H) spherical shape with 20 faces
9. lysogenic induction	(I) smallest unit of viruses to cause an effect
10. oncogene	(J) codes for protein to bring about transformation

FILL IN

For the following graph describe the actions occurring at times A, B, C and D.

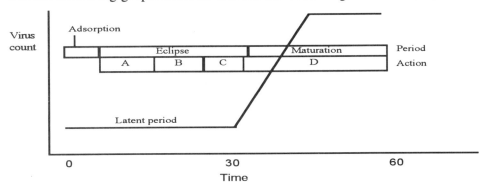

MULTIPLE CHOICE

1. Arrange the viral infection process into the appropriate order.

    1. Replication   3. Attachment

    2. Assembly of particles   4. Penetration

    (A) 3,1,4,2 (B) 3,4,1,2 (C) 1,3,2,4 (D) 4,3,2,1 (E) 4,3,1,2

2. Which of the following processes does NOT occur during viral infection of any host cell?

    (A) attachment and penetration

    (B) cytopathic effect on host cell

    (C) viral multiplication by binary fission

    (D) release from the host cell

    (E) uncoating

71

3. The form of a bacterial virus which is covalently inserted into its host's DNA is called a/an:

(A) bacteriophage

(D) virulent phage

(B) prophage

(E) retrovirus

(C) temperate phage

4. Genes from a lytic bacterial virus may encode all of the following EXCEPT:

(A) capsid proteins

(B) lysozyme

(C) repressors which prevent expression of early genes

(D) proteins involved in takeover of cell metabolism

(E) proteins involved with viral genome replication

5. The specificity of viruses for certain types of cells is usually explained by events which occur in which of the following phases of the virus replication cycle?

(A) attachment

(D) assembly

(B) penetration

(E) release

(C) replication of virus components

6. An *E. coli* cell which carries a lambda prophage is immune to a lytic infection by a second lambda virus because:

(A) the second virus cannot adsorb to the cell

(B) the second virus cannot inject its DNA

(C) proteins from the prophage's lytic genetic program prevent replication of the second virus

(D) proteins from the prophage's lysogenic genetic program prevent replication of the second virus

(E) the bacterium has already been lysed

7. Animal viruses enter cells by

(A) attachment to pili

(D) endocytosis

(B) budding

(E) diffusion

(C) holes in the cell membrane

8. Tumor virus(es)

1. can have either RNA or DNA genomes

2. "transform" the cells they infect

3. cause cells to divide without control

4. can cause malignancies in animals

5. genomes are often integrated into the host DNA

(A) 3 (B) 1,3 (C) 2,4 (D) 1,2,3,4,5 (E) 1,2,4,5

9. Viruses which possess reverse transcriptase will be:

    (A) DNA viruses

    (B) RNA viruses

    (C) DNA-RNA hybrid viruses

    (D) A and C only

    (E) A, B, and C

10. The genome of rhabdoviruses consists of a single-stranded RNA molecule whose sequence is complementary to the RNA sequence which functions as a messenger RNA. How is the "+ sense" messenger RNA produced in cells infected by rhabdovirus?

    (A) reverse transcriptase activity;

    (B) host cell RNA polymerase activity

    (C) one portion of the infecting RNA is directly translated by host cell ribosomes

    (D) the infecting virus particle contains an RNA-dependent RNA polymerase

11. Gene sequences related to oncogenes have been found in all of the following EXCEPT:

    (A) transformed cells

    (B) malignant tumor cells

    (C) normal cells

    (D) benign tumor cells

    (E) The sequences have been found in all of the above

12. Lysogenic induction describes the process where cells are induced to produce virus and burst. This process requires:

    (A) transformed cells

    (B) a lysogenic bacteria and an induction agent

    (C) a lysogenic bacteria and lytic virus

    (D) lambda repressors

    (E) bacteriophage Mu

13. Rhabdovirus is an important human pathogen. It is different because:

    (A) it is rod shaped and a negative-stranded RNA virus

    (B) it is a lysogenic virus with its own induction agent

    (C) it has a complex lipid envelope

    (D) it is an mRNA repressor

    (E) two of the above are correct

14. Viroids are typical of other viruses EXCEPT:

    (A) they lack any proteins

    (B) they have circular single-stranded RNA molecules

    (C) they are the smallest known pathogen

    (D) they have no means to code for enzyme production

    (E) all of the above describe viroids

15. Provirus describes:

    (A) a virus whose replication has a step where DNA is formed from an RNA genome

    (B) the genome of a temperate virus when it is integrated into and replicating with a host genome

    (C) a class of virus that infects only prokaryotic organisms

    (D) a zone of clearing

    (E) none of the above

DISCUSSION

1. Viruses lack enzymes for catabolic or anabolic metabolism, yet they are able to replicate. How do they accomplish this?

2. In general, how does the genetic material of viruses differ from that of cells? What problems does this entail in terms of genome replication?

3. Compare and contrast the lytic and lysogenic cycles of bacteriophage replication in terms of gene expression, increase in virus numbers, and maintenance in the environment.

4. What techniques are used to quantify the number of virus particles in a sample?

5. Give three examples of how an infecting virus shuts off the expression of the host's genetic material.

6. By what mechanism(s) are "early" viral genes expressed in the case of (a) double-stranded DNA viruses, (b) single-stranded DNA viruses, (c) + sense single stranded RNA viruses, and (d) - sense single stranded RNA viruses?

7. Why is it necessary for some viruses to contain enzymes in the virus particle?

8. Compare the processes by which bacteriophages penetrate cells with those of animal viruses.

9. What four proteins are encoded by the poliovirus genome, and what is the function of each?

10. Several viruses have genomes in which the ends have complementary base sequences. What purposes do these complementary or repeat sequences have?

11. What is cancer? What role can viruses have in causing cancers? By what mechanism do viruses transform cells?

ANSWERS

Completion

1. capsomeres, self-assembly, helical, icosahedral; 2. cancer; 3. plaque; 4. eclipse; 5. phagocytosis; 6. host; 7. host-range; 8. phenotypic mixing; 9. +; 10. double; 11. late; 12. hydroxymethylcytosine, glucosyl residues; 13. repressor protein; 14. mu; 15. envelopes, cell membrane; 16. poliovirus; 17. negative, RNA; 18. papovavirus, retrovirus; 19. poxvirus; 20. oncogenes.

Multiple choice

1. B; 2. C; 3. B; 4. C; 5. A; 6. D; 7. D; 8. D; 9. B; 10. D; 11. E; 12. B; 13. E (a&c); 14. E; 15. B.

Matching

1. H; 2. C; 3. B; 4. G; 5. I; 6. F; 7. D; 8. A; 9. E; 10. J.

Fill in

A. early enzymes are formed by the virus; B. nucleic materials formed; C. protein coat formed; D. virus assembled and escapes from cell.

Discussion

1. See text Section 8.5
2. See Chapter Introduction and text Sections 8.2 and 8.7
3. See text Sections 8.5, 8.6, and 8.12
4. See text Sections 8.3 and 8.4
5. See text Sections 8.11, 8.12, and 8.17
6. See text sections 8.8, 8.9, 8.10, 8.15, and 8.16
7. See Chapter introduction and text Section 8.6
8. See text Section 8.6
9. See text Section 8.15
10. See text Sections 8.11 through 8.13 and 8.22
11. See text Sections 8.3, 8.14, 8.21, and 8.22

MICROBIAL GENETICS

OVERVIEW

Chapter 9 (pages 306-359) focuses on how the genetic information can be changed: either by mutation or by the transfer of genes from one organism to another. The successful transfer of genetic information includes two elements -- the introduction of genes from a **donor** cell into a **recipient** and **recombination** of those introduced genes into the recipient's genome. Microbial genetics are important because the genes are the are the basis for cell function and microorganisms are excellent tools for studying gene function.

CHAPTER NOTES

Mutations are inheritable changes in the base sequence of nucleic acid -- the genetic material. An organism with these changes is called a **mutant**. **Genetic recombination** is the process where genes from two genomes are combined together. A mutant will be different from its parent, its **genotype** or genetic makeup has been altered. The phenotype or visible properties of the mutant may or may not be altered. The genotype of a strain is indicated by use of three small italics letters followed by a capital letter and indicates the gene involved in a process (*his*C indicates the gene for HisC protein). The phenotype of the strain is indicated by three letter code that ends in a +/-. For example Thr$^+$indicates a strain can make its own threonine while Thr$^-$ indicates that it cannot. An **auxotroph** is formed when a required nutritional material (amino acid for example) that the parent strain, **prototroph**, could make is no longer formed.

Mutations can occur **spontaneously**, because of mistakes during replication or due to natural radiation at a frequency of about one in 1,000,000, or may be experimentally induced using **mutagens**. Mutations can be chemically induced by **base analogs**, compounds that are structurally similar to the purines and pyrimidines in DNA. The cell incorporates them into DNA, but during subsequent replication, the analogs have a higher probability of base pairing incorrectly, thereby inserting the wrong base into the new DNA strand. Other chemical mutagens react directly with DNA to alter the bases. **UV radiation** is absorbed by the purines and pyrimidines in DNA, and one of its effects is to form **pyrimidine dimers** in one strand, which prevents these thymine bases from pairing correctly during replication. **Ionizing radiation** generates free radicals in cells, and these can react with the DNA backbone to cause breaks. Biological agents, such as **transposons** and the bacteriophage **Mu**, cause mutations by inserting DNA sequences into genes, and thereby disrupting the coding information.

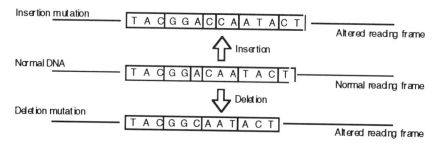

A mutation may be a change in a single base pair (**point mutation**) or involve large **deletions** or **insertions** of base pairs. The insertion of a single additional base into a gene can have dramatic effects upon the amino acid sequence of the protein produced from that gene, due to the **reading-frame shift** this causes in the translation of the mRNA produced from the gene. It is possible to move large sections of DNA to a second location and the process is termed **translocation.** If the mutated gene is part of an operon (see Section 5.10) the mutation may exert **polar** effects upon other genes in the operon. The effects of a specific mutation may be reversed by a second (**suppressor**) mutation in either the same gene, or in another gene. Note that cells do have **DNA repair systems** to correct damage to DNA. The **SOS system** is one of these, but it is **error-prone** and the repaired DNA may still contain mutations.

One use of mutant bacterial strains has been to determine the potential mutagenicity of chemicals -- either manufactured or natural. The **Ames test** utilizes back mutation in a strain of bacteria that are auxotrophic for a nutrient. When auxotrophic cells (His$^-$) are spread on a medium that lacks histidine no growth will occur. If, however, the cells are treated with a chemical that causes a reversion mutation it can then grow.

General or **homologous recombination** requires extensive **homology** and is mediated by an enzyme, **RecA protein.** The sequence of events are (1) nicking of a DNA molecule, (2) opening of the DNA double helix, (3) pairing between homologous single strands of two DNA molecules (requires presence of RecA), and (4) breakage and rejoining of DNA strands so that portions of the DNA molecule are exchanged. An important point is that this process leads to new genotypes only if the two molecules that are recombining differ genetically in regions outside those where breakage and rejoining occurred. In order to detect recombination or exchange of DNA, the offspring must be phenotypically different from the parent.

In bacteria, the **gene transfer** that precedes recombination can occur by three mechanisms: transformation, transduction, or conjugation. **Transformation** involves the uptake by a recipient

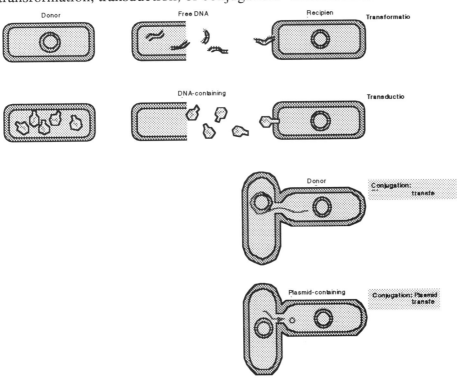

of free or naked DNA released from a donor. However, cells may only be physiologically **competent** to take up DNA. Competence is related to changes in the cell surface that allow strong binding of DNA. In some organisms, such as *E. coli,* the transformation process can be enhanced by special pre-treatment of cells. The cell can undergo **electroporation** where small holes or pores are open in the cell. A single strand of the transforming DNA is integrated into the chromosome, using general recombination mechanisms. A cell with a new genotype is generated when this strand is replicated and the resulting molecule forms the genome of a new cell, at cell division. Eukaryotic cells can also be treated to take up free DNA, although the specific treatments are different from those used in bacteria.

In **transduction**, the transferred DNA is carried in the capsid of a bacteriophage. The donor's DNA replaces part or all of the viral genome in the phage head. Thus, the particle is probably **defective** in viral replication because essential viral genes are missing. In the case of temperate phages such as lambda, bacterial DNA becomes associated with the virus genome when the prophage is excised incorrectly from the bacterial chromosome. When this occurs, the same set of bacterial genes is always incorporated into lambda phage. This phenomenon is **specialized transduction**, because it is only effective in transducing a few special bacterial genes. In contrast, **generalized transduction** can transfer any bacterial gene to the recipient. This process may occur with phages that degrade their host DNA into pieces the size of viral genomes. If these pieces are erroneously packaged into phage particles, they can be delivered to another bacterium in the next phage infection cycle. Phages P22 of *Salmonella typhimurium* and P1 and mu of *E. coli* carry out generalized transduction.

Some temperate phages cause phenotypic changes in the bacteria they infect even without transducing bacterial genes. In the lysogenic state, viral genes are expressed which confer new properties on the cell. Examples of this **phage conversion** are toxin production by pathogenic bacteria such as *Corynebacterium diphtheriae* and surface polysaccharide structure in *Salmonella anatum*.

Conjugation, the third means of gene transfer is mediated by special genetic elements called plasmids. **Plasmids** are defined as small, circular DNA molecules that reproduce autonomously. While plasmids are DNA, they control their own replication separately from that of the chromosome. The presence of plasmids in cells can be detected by techniques that separate them from chromosomal DNA. This involves buoyant density differences due to the tight **supercoiling** of these rather small DNA circles; the density difference can be enhanced by adding compounds that **intercalate** between DNA base pairs, such as **ethidium bromide**. The tightly wound plasmid DNA cannot bind as much ethidium bromide as the chromosomal fragments. Adding ethidium bromide, or other treatments that affect DNA, to whole cells at the appropriate concentration may **cure** cells of their plasmids. If plasmid replication is more sensitive to these agents than chromosome replication, plasmids may not segregate to all progeny cells during cell division (see Figure 7.18).

Some (but not all) have genes that can direct their transmission from one cell to another by conjugation. Finally, plasmids may have genes that confer novel phenotypes on cells, such as resistance to antibiotics, production of toxins, or the capacity to metabolize unusual substrates such as pesticides or industrial solvents. Antibiotic resistance is conferred by R plasmids. These plasmids have diminished the effectiveness of antibiotics in combating infectious diseases because (i) they may confer resistance to as many as five different antibiotics at once upon the cell, and (ii) by conjugation, they can be rapidly disseminated through the bacterial population. Multiple antibiotic resistance is a consequence of their construction -- they contain several **transposons**, each of which confers resistance to a unique antibiotic. The

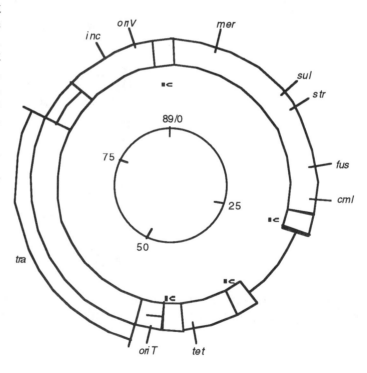

genes in the transposons generally specify an enzyme that **inactivates** the drug before it enters the cell and reaches its target. This differs from chromosomal mutations that result in antibiotic resistance. These generally are modifications of the antibiotic's **target** of action.

Plasmids are autonomously replicating molecules. What elements are necessary to control DNA replication? There must be an **origin of replication**, where the frequency of replication can be regulated. The number of plasmid copies is tightly regulated at a few copies with some plasmids, whereas in others, initiation of replication is relatively uncontrolled, and twenty to thirty plasmid copies may be present in a cell. In general, the enzymes used for DNA replication are those coded by the chromosome -- it is the regulatory genes that are plasmid encoded.

Conjugative plasmids initiate gene transfer by altering the cell surface to allow contact between the plasmid-containing donor cell and a plasmid-less recipient. A plasmid gene codes for the production of a **sex pilus** that initiates pair formation. Subsequently, a **conjugation bridge** is formed through which DNA is transferred. The transfer of plasmid DNA is accompanied by its replication. That is, the donor cell does not lose its plasmid but transfers a copy to the recipient. In actual fact, replication is shared between donor and recipient. A single DNA strand is transferred as a consequence of **rolling circle replication** in the donor; this strand is used as a template by the recipient to generate a double stranded DNA molecule. Therefore, the consequence of conjugation is that both the donor and the recipient cells contain the plasmid. The recipient is now competent to serve as a plasmid donor in other conjugations.

Some conjugative plasmids, such as the F factor in *E. coli*, can also direct transfer of chromosomal genes by conjugation. *E. coli* strains which have this property are **Hfr strains**. The F factor can integrate into the chromosome to form one DNA molecule. This occurs at regions of homology between F and the chromosome. These regions are **insertion sequences** located on both molecules. F factor can now transfer chromosomal genes during a conjugation, because in effect, the chromosome has become part of the F factor. It is the F factor that has the genetic information to drive gene transfer. Specifically, there is a nucleotide sequence on F that specifies the **origin of transfer**. The host chromosome was inserted just downstream from this region. DNA is transferred just as described above for plasmid transfer. It is important to note that chromosomal genes are transferred before any of the plasmid genes. Thus, if the cytoplasmic bridge is broken before the entire chromosome is transferred, the recipient remains F⁻.

Just as improper excision of lambda prophage leads to high frequency transfer of some chromosomal genes by specialized transduction, the improper excision of F factor results in the F factor containing a few chromosomal genes. Conjugation involving such **F' factors** transfers these particular chromosomal genes at high frequency.

Irrespective of their ability to drive conjugation, plasmids confer interesting properties on cells. We have said that conjugative plasmids encode the production of pili. Note that these **pili** may make the cell susceptible to phage infection, as some RNA phages use these pili as receptors.

A cell can contain several different plasmids. However, some plasmids are **incompatible** with each other. That is, they cannot be maintained together in a single cell. This may be a consequence of similarity between the regulatory elements that control plasmid replication.

Conjugation with Hfr strains was used to map the *E. coli* chromosome. Genes are transferred in a linear, sequential process. Conjugants were deliberately broken apart at different time intervals, and the recipients were analyzed for the genes they received. The shorter the time interval of the interrupted mating in which two genes were successfully co-transferred, the closer they must physically be located on the chromosome.

Transposons and **insertion sequences** are genetic elements capable of moving within cells. Transposons differ from insertion sequences in that they contain additional genes, such as ones for antibiotic resistance. The frequency with which these elements move is rather low, but 10 to 100 fold greater than the frequency of spontaneous mutation. The ends of these elements contain **repeated sequences**. In addition, they code for a **transposase** enzyme that can insert the elements at any point into a DNA molecule.

When these transposable elements insert into a DNA target sequence, that target sequence is duplicated. In addition, elements that undergo **replicative transposition** also are duplicated. That is, a copy remains at the original site, and the other copy is inserted at a new site. The transposase makes single strand cuts in the inverted repeat sequences at the ends of the transposable element, and at the target site. The element is joined to the target via the single strand ends, and the gaps are filled in by DNA replication. Finally, the **cointegrate** formed by recombination is resolved to generate a copy of the transposable element at the new site. In other transposons (such as **Tn5**), transposition is **conservative**, and the transposon is excised from its original location, and is reinserted at a new site. If the site of transposon insertion is within an existing bacterial gene, it is likely to be inactivated, and a mutation has occurred.

Another type of genetic rearrangement is the **inversion** of a DNA segment in the genome. This has regulatory significance if the segment contains a promoter. Remember that the orientation of the promoter determines the direction in which mRNA synthesis will occur. Therefore, inverting a promoter sequence leads to expression of a different gene. This mechanism is responsible for changes in the type of flagellar protein produced by *Salmonella*.

The genetic transfer mechanisms discussed in this chapter have been used to map the order of genes on bacterial chromosomes. **Interrupted mating** was useful in coarsely determining gene order, because large DNA segments can be transferred in conjugation. To accurately determine the order of closely linked genes, transduction has been most useful, because small DNA pieces are transferred.

The three genetic exchange mechanisms described above can and have been used to map the location of genes in the bacterial chromosome. In *E. coli*, the location of some 1900 genes has been identified. This type of mapping has revealed that the genes that control a given pathway are often clustered or closely linked on the bacterial chromosome. This grouping of genes has lead to the **operon concept** that theorizes that the closely linked genes are under a common control mechanism.

In eukaryotic organisms, new combinations of genes are assembled on a regular basis, as a consequence of sexual reproduction. Eukaryotic microorganisms undergo an **alternation of generations**, in which the number of chromosomes per cell varies by a factor of two. At some point, the **haploid** cells, which contain one copy of each chromosome, function as gametes and fuse to form a **diploid** zygote. In this zygote, new gene combinations can arise from the mixture of the gametes' genomes.

Gametes are generated by **meiosis**. A diploid cell divides into 2 cells without replicating its chromosomes -- each chromosome pair the cell possesses is divided between the two daughter cells. Thus, they are now haploid. These cells replicate their chromosomes and divide again, so that four gametes are produced from the diploid cell.

More is known about the genetics of *Saccharomyces cerevisiae* (yeast) than of other eukaryotes. For most of their life cycle, yeasts are haploid. These cells can serve as gametes, if cells of the two different **mating types** fuse. Mating types differ in the **hormones** they excrete, and the **receptors** on their surface that interact with the complementary hormone. The resulting diploid zygote undergoes meiosis to form four haploid **ascospores**, each of which can germinate to form a vegetative cell. Yeast genetics is attractive because the four ascospores can be isolated and germinated separately. Thus, the consequences of sexual reproduction can be unequivocally determined.

The mating type of yeast cells is regulated by an insertion of DNA segments, called the **cassette mechanism**. The DNA sequence transcribed from the **MAT promoter** can be replaced with a copy of the gene of one of the two mating types located elsewhere on the yeast genome. These genes are inactive unless copied to the "reading head", the MAT promoter.

Eukaryotes contain DNA not only in the nucleus but also in two organelles -- mitochondria and chloroplasts. Although most of the genetic information to produce these structures is in the nucleus, there are a few functions retained in the organelles' DNA. For progeny cells to obtain these organelles, the existing structures must divide. Thus, the genetic information in these structures is inherited separately from that in the nucleus. This information does not code for many proteins, but does specify the ribosomal and transfer RNA used for protein synthesis within the organelles. A unique feature of protein synthesis in mitochondria is that some mRNA codons are translated differently than in all other systems.

SELF TESTS

COMPLETION

1. A genotypic change resulting from the incorporation of free DNA originating from another cell is _____.

2. Bacterial cells that can bind and incorporate free DNA molecules are in a state of _____.

3. The process of transferring genetic information directly between cells in contact with one another is _____.

4. A pilus is required for genetic exchange by _____.

5. *E. coli* clones that efficiently transfer a variety of chromosomal genes are _____ strains.

6. In a mating between an Hfr cell and a F⁻, the result is that the donor cell is _____ and the recipient is _____.

7. In a mating between an F' cell and a F⁻, the result is that the donor cell is _____ and the recipient is _____.

8. The transfer of bacterial DNA between cells mediated by a bacteriophage is called _____.

9. If a prophage is excised incorrectly from the chromosome, genetic transfer by _____ may occur.

10. Phages that can transfer any part of the bacterial genome to another cell are _____ transducing phages.

11. The largest amount of DNA can be transferred between bacteria by the process of _____.

12. General recombination involves _____ protein.

13. Site-specific recombination requires a _____ protein and short, specific _____.

14. Viruses that can attach to and penetrate host cells but are unable to replicate are _____.

15. The enzymes used to replicate plasmid DNA are encoded on the _____ whereas the proteins that regulate the timing of replication are encoded on the _____.

16. The proteins produced from plasmid-encoded antibiotic resistance genes are _____.

17. Transposition events occur at frequencies of _____.

18. A diploid cell is formed by the fusion of two haploid _____.

81

19. If a diploid cell contains two identical copies of a particular gene, then it is _____ for that trait.

20. A haploid yeast cell of "a" mating type can form diploids with cells of _____ mating type.

21. A single base change in a gene may alter a large number of amino acids in the protein if it is a _____ mutation.

22. Ultraviolet radiation forms _____ in DNA. These can be repaired after exposure to visible light by _____.

KEY WORDS AND PHRASES

allele (p 351)	Ames test (p 317)
auxotroph (p 309)	bacteriocin (p 335)
carcinogen (p 317)	cis (p 321)
colicin (p 335)	competence (p 324)
complementation test (p 320)	conjugation (p 335)
conjugative plasmids (p 332)	cytoplasmic inheritance (p 351)
diploid (p 351)	DNA inversion (p 344)
dominant allele (p 351)	donor (p 335)
electroporation (p 326)	enterotoxin (p 334)
F plasmid (p 333)	F^+ (p 336)
F^- (p 336)	frame-shift mutation (p 312)
genetic map (p 339)	genetic marker (p 307)
genotype (p 307)	haploid (p 351)
helper phage (p 329)	heterozygous (p 333)
homozygous (p 333)	inverted terminal repeat (p 341)
lambda dgal (p 329)	mating (p 335)
nick (p 318)	nonconjugative plasmids (p 332)
penicillin selection (p 309)	phage conversion (p 330)
phase variation (p 344)	phenotype (p 307)
plasmid curing (p 332)	prototroph (p 309)
recessive allele (p 351)	recipient (p 336)
replica plating (p 309)	reversions (p 312)
rolling circle replication (p 331)	silent mutation (p 312)
Tn10 (p 342)	Tn5 (p 342)
trans (p 321)	transduction (p 327)
transfection (animal cells) (p 326)	transfection (bacteria) (p 325)
transformation (p 322)	transposase (p 342)

transposition (p 341)	transposon mutagenesis (p 342)
wild type (p 309)	

MULTIPLE CHOICE

1. What features are common to transformation, transduction, and conjugation?

    1. transfer of genetic information is unidirectional.

    2. there is a fusion of two haploid chromosomes.

    3. recombination occurs by breakage and reunion.

    4. gene transfer is usually incomplete.

    5. following recombination, the recipient undergoes a meiotic division.

    (A) 1,3,4 (B) 2,5 (C) 1,3 (D) 3,4 (E) 1,2,3,4,5

2. Gene transfer in bacteria by transformation has the following characteristic:

    (A) a majority of the donor genes are transferred.

    (B) it involves a plasmid.

    (C) it depends on phage infection of the recipient cell.

    (D) it can be carried out using free DNA extracted from the donor.

    (E) all bacterial species can carry out the process.

3. After β phage infects *Corynebacterium diphtheriae*, the bacterium is capable of causing diphtheria. Which term is the **best** description of this phenomenon?

    (A) lysogeny   (D) phage conversion

    (B) transformation   (E) general recombination

    (C) transduction

4. Transformation does not involve

    (A) uptake of DNA

    (B) competent cells

    (C) the restriction-modification system

    (D) cell to cell contact

    (E) physical substitution of part of the recipient DNA by donor DNA

5. Which of the following does not apply to conjugation?

    (A) There is a one-way transfer of genetic information.

    (B) Requires cell-to-cell contact.

    (C) Occurs after the induction of a lysogenic cell.

    (D) In *E. coli*, the F factor can direct the process.

    (E) Can be used to transfer chromosomal DNA.

6. Which of the following statements is false concerning a mating between F$^+$ and F$^-$ cell?

   (A) The F$^-$ cell is converted to an F$^+$ cell.

   (B) The F$^+$ cell is converted to an F$^-$ cell.

   (C) Chromosomal genes are rarely transferred.

   (D) Cell-to-cell contact is always necessary.

   (E) The genes involved in pilus formation are transferred at high frequency.

7. In an Hfr strain

   (A) the F factor has integrated into the chromosome.

   (B) a high frequency of recombination is occurring because transposons are "jumping" between DNA molecules.

   (C) the F factor is transferred first in conjugation.

   (D) the plasmid involved in conjugation is replicating autonomously.

   (E) lambda prophage directs transfer of chromosomal genes.

8. R plasmids are medically important because they

   (A) cause certain types of bacterial diseases.

   (B) carry genes specifying resistance to certain antibiotics.

   (C) encode for enterotoxin production.

   (D) convert non-pathogenic bacteria to pathogens.

   (E) can transform eukaryotic cells into tumor cells.

9. Bacterial conjugation

   (A) is common among bacterial species

   (B) requires cell to cell contact which is initiated by the sex pilus

   (C) usually results in transfer of all the donor chromosome

   (D) usually converts the recipient to an Hfr

10. All of the following are true of conjugation in *E. coli* except:

   (A) it requires a plasmid

   (B) it always results in transmission of plasmid DNA to the recipient

   (C) it may not involve transfer of a complete chromosome from donor to recipient

   (D) it can result in transfer of the complete chromosome from donor to recipient

   (E) it requires a sex pilus

11. In conjugation, genes on the bacterial chromosome are most frequently transferred in matings between:

   (A) two F$^-$ strains   (D) B and C above

   (B) F$^+$ and F$^-$ cells   (E) all of the above

   (C) Hfr and F$^-$ cells

84

12. Among the following the relationship most similar to lysogeny is

    (A) F$^+$ state   (D) Hfr state

    (B) F$^-$ state   (E) spore state

    (C) F$^'$ state

13. Interrupted mating experiments between Hfr and F$^-$ strains of *E. coli* have led to the following conclusions:

    1. gene transfer is in a temporal order from F$^-$ to Hfr.

    2. gene transfer is in a temporal order from Hfr to F$^-$.

    3. genes can be mapped by their time of entry.

    4. all gene transfer occurs within 5 minutes of contact.

    5. both strands of DNA are transferred during conjugation.

    (A) 1,2 (B) 2,3 (C) 3,4 (D) 4,5 (E) 1,5

14. The F factor and lambda are

    (A) nonessential for host cell viability.

    (B) lethal to the host.

    (C) only replicated autonomously.

    (D) replicated after recombination with the chromosome.

    (E) carried only by competent cells.

15. Plasmids may be functionally involved in all of the following except:

    (A) conjugation   (D) transduction

    (B) transfer of chromosomal genes(E) transfer of drug resistance

    (C) synthesis of pili

16. When a bacteriophage transfers bacterial DNA from one cell to another:

    (A) the recipient cell dies in the process.

    (B) the phage may have acquired the DNA as a consequence of lysogeny.

    (C) both cells must be in close proximity to one another.

    (D) the phage acquired the DNA by transformation.

    (E) the donor cell undergoes phage conversion.

17. A transducing phage:

    (A) contains only viral DNA

    (B) may contain viral and bacterial DNA

    (C) is sensitive to DNAase

    (D) can never transfer extrachromosomal genes

    (E) contains one or more transposons

18. The gene LEAST likely to be transferred would be one on a

  (A) F$^+$

  (B) F$'$

  (C) R factor

  (D) chromosome

  (E) mu phage

19. The exchange of homologous DNA molecules in general recombination requires

  (A) transposons

  (B) insertion sequences

  (C) the breakage and reunion of DNA strands

  (D) reverse transcriptase

  (E) low levels of DNA homology

20. Plasmids can be separated from chromosomal DNA because there are differences in

  (A) nucleotide sequence

  (B) buoyant density

  (C) the types of nucleotides they contain

  (D) cellular location

  (E) size

21. Which of the following processes does not involve DNA synthesis?

  (A) conjugation

  (B) transposon mutagenesis

  (C) general recombination

  (D) meiosis

  (E) mitosis

22. Which of the following statements is FALSE?

  (A) The bacterial chromosome is comprised of smaller units called genes.

  (B) The genotype of an organism is defined as the sum total of the genetic potential of the cell.

  (C) The phenotype of an organism can be environmentally influenced.

  (D) The function of many regulatory mechanisms is to insure efficient use of carbon and energy.

  (E) Spontaneous mutations always occur in bacteria at a rate of 10^{-6}.

23. An *E. coli* strain that requires the amino acid lysine for growth is termed a/an

  (A) autotroph

  (B) aminotroph

  (C) auxotroph

  (D) prototype

  (E) prototroph

24. The Ames test is a simple and cost-effective method to screen chemical agents that might be carcinogenic. The rationale behind the test is that

  (A) the rate of spontaneous mutation in bacteria is much higher than in eukaryotic cells.

  (B) bacteria have no repair mechanisms to cope with alterations in their DNA.

  (C) mutations in bacteria always result in auxotrophy.

  (D) most carcinogenic agents are mutagenic.

  (E) bacteria are transformed to tumor cells more easily than are eukaryotic cells.

MATCHING

Match the term with its definition.

1. auxotroph	(A) able to take-up DNA
2. genotype	(B) surface structure for pair formation
3. suppressor mutations	(C) chemical mutagens resembling DNA base
4. base analogs	(D) compensate for a primary mutation
5. transduction	(E) rod shaped
6. transformation	(F) free DNA, entry into cell
7. conjugative plasmid	(G) plasmids that control their own transfer
8. curing	(H) change from diploid to haploid state
9. R plasmid	(I) agents to kill closely related species
10. bacteriocins	(J) antibiotic resistance agent
11. sex pilus	(K) nucleotide sequence in DNA
12. meiosis	(L) inhibition of plasmid replication
13. Bacterium	(M) virus mediated entry into cell
14. transposons	(N) DNA base sequence which controls its own movement (Tn)
15. competent	(O) a nutritional mutant

DISCUSSION

1. What similarities are there in bacterial genetic exchange mediated by transformation, transduction, and conjugation?

2. Contrast transformation, transduction, and conjugation with respect to (a) the maximum number of genes transferred, and (b) the proteins required to effect gene transfer.

3. Compare the base-pairing requirements and proteins involved in general and site-specific recombination.

4. What steps involving the breakage of the sugar-phosphate backbone of DNA are a part of general recombination? How are these breaks repaired?

5. What characteristics do cells competent for transformation have? How can these properties be enhanced?

6. What characteristics of a phage's life cycle would make it a candidate for a specialized transducing phage, or a generalized transducing phage?

7. What type of genes are carried on all plasmids? What genes are required to make a plasmid conjugative?

8. By what molecular mechanism are Hfr strains formed?

9. Describe the molecular events involved in the transposition of an element like Tn5.

10. What similarities and differences are there between meiosis and mitosis?

11. At what stages of its life cycle is a yeast cell haploid? diploid?

12. Will all point mutations in a gene lead to an observable change in the protein for which it codes? Consider (a) the degeneracy of the genetic code, (b) the chemical similarity of some amino acids to one another (for example, leucine and isoleucine), and (c) the location of the alteration in relation to the active site of the enzyme.

13. What effect is the deletion or insertion of one base in a gene likely to have upon the amino acid sequence of the specified protein?

14. What similarities and differences are there between insertion sequences and transposons?

ANSWERS

Completion

1. transformation; 2. competence; 3. conjugation; 4. conjugation; 5. Hfr; 6. Hfr, F$^-$; 7. F', F$^{'}$; 8. transduction; 9. specialized transduction; 10. generalized; 11. conjugation; 12. RecA; 13. transposase, nucleotide sequence; 14. defective; 15. chromosome, plasmid; 16. inactivating enzyme; 17. 10^{-5} - 10^{-7}; 18. gametes; 19. homozygous; 20. alpha; 21. polar; 22. thymidine dimers, photoreactivation.

Multiple choice

1. A; 2. D; 3. D; 4. D; 5. C; 6. B; 7. A; 8. B; 9. B; 10. B; 11. C; 12. D; 13. B; 14. A; 15. D; 16. B; 17. B; 18. D; 19. C; 20. B; 21. C; 22. E; 23. C; 24. D.

Matching

1. O; 2. K; 3. D; 4. C; 5. M; 6. F; 7. G; 8. L; 9. J; 10. I; 11. B; 12. H; 13. E; 14. N; 15. A.

Discussion

1. See text Sections 9.6, 9.7, and 9.9.   .

2. See text Sections 9.6, 9.7, and 9.9.   .

3. See text Section 9.5.

4. See text Section 9.5.

5. See text Section 9.6.

6. See text Section 9.7.

7. See text Section 9.8.

8. See text Section 9.9.

9. See text Section 9.11.

10. See text Section 9.12

11. See text Section 9.13.

12. See text Section 9.8.

13. See text Section 9.2
14. See text Section 9.11.

GENETIC ENGINEERING AND BIOTECHNOLOGY

OVERVIEW

Chapter 10 (pages 359-398) is concerned with **genetic engineering** that involves a set of sophisticated techniques in which particular genes of interest are isolated, their DNA sequences modified, and the regulation of their expression altered. Genetic engineering has application in both research and applied systems. Genetic engineering used in commercial applications is often referred to as **biotechnology**.

CHAPTER NOTES

The objective of a genetic engineering program is to isolate and purify specific genes in a process known as **gene cloning**. The underlying strategy of cloning is to isolate and recover a specific gene from a large, complex genome and insert it into a small, easily manipulated one. Total DNA is isolated from an organism, and it is cut into smaller pieces using a specific **restriction endonuclease** (also see Text section 6.3). Recall that restriction enzymes make a staggered cut through a double-stranded DNA molecule at a specific sequence or palindrome. Thus, the ends of cut molecules have complementary single-stranded sequences. If this preparation of **restriction fragments** is mixed together with a **cloning vector** that has been cut with the same restriction enzyme, hybrid molecules are formed in which a fragment of source DNA is inserted into the cloning vector's DNA. The mixture of **recombinant** molecules is introduced into a host bacterium to generate a

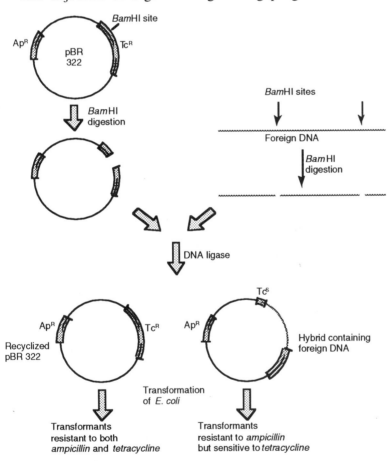

DNA library. That is, individual bacterial clones contain different recombinant DNA molecules. If one examined enough clones, the probability is great that you would find all of the original restriction fragments in the source DNA represented.

The cloning vectors play an important role in this process, because they must constitute a delivery system to introduce the recombinant molecules formed in a test tube into cells, and must

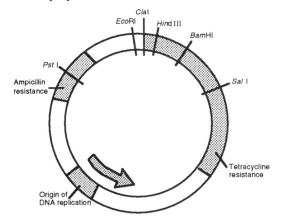

be capable of autonomous replication to maintain the recombinant DNA in the cell. Also, subsequent analysis is easier if the vector is a small DNA molecule. Plasmids and bacteriophage have been the most useful cloning vectors.

Recombinant plasmids can be introduced into cells by transformation. After introduction, they replicate autonomously, and with some plasmids, many plasmid copies are present in a cell, which of course increases the amount of the recombinant DNA per cell. Furthermore, plasmids are useful vectors because they can contain **selectable genetic markers,** such as antibiotic resistance genes, which can be used to select only those host cells that have received the plasmid during transformation.

Another desirable characteristic for a plasmid cloning vector is its small size, because this increases the amount of foreign DNA that can be inserted. There is a limit to the size of DNA that can transform cells; the bigger the plasmid, the less room there is for foreign DNA.

The cloning vector should only have one site for a particular restriction enzyme. That way, the location of recombinant DNA within the molecule is precisely known. Another useful characteristic is that this site is within an antibiotic resistance gene. Then, the experimenter can easily determine which host clones have recombinant plasmids, because these will not be resistant to the particular antibiotic. The foreign DNA has made the resistance gene nonfunctional by **insertional inactivation**.

Bacteriophage lambda has been a useful cloning vector. One third of its genome can be removed and substituted with foreign DNA. This is done in vitro, again using restriction enzymes. DNA is then packaged into phage particles in vitro. These intact phage are delivery systems to infect *E. coli* with recombinant DNA in the lambda genome. Wild type lambda is not used in cloning because they contain too many restriction sites. A modified lambda with fewer restriction sites is used. When the lambda is modified to contain only two

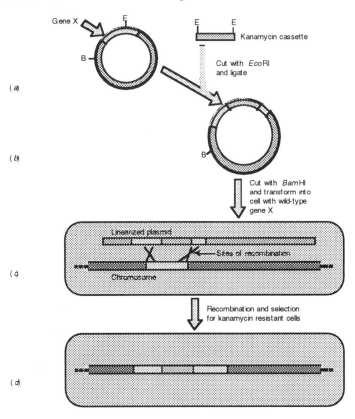

restriction sites, it is possible to remove this section of DNA and replace it with foreign DNA. This system is referred to as a **replacement vector** and allows cloning of large DNA fragments.

Cosmids are vectors that combine the advantages of plasmids and phage lambda. They are plasmids into which the *cos* site of lambda has been cloned. The *cos* **site** allows the recombinant

91

plasmid to be packaged into lambda virions. The recombinant plasmid can be introduced into *E. coli* more efficiently than by transformation, and can be stored more stably in phage particles.

The **Human Genome Project**, an effort to map and sequence the entire human genome, has resulted in the development of the **yeast artificial chromosome** (YAC). The YAC vector is about 10 kb, can accept 200 to 800 kilobase pairs of DNA and has been designed to replicate in yeast like a normal eukaryotic chromosome.

The **bacteriophage M13** was made into a useful cloning vector for DNA sequencing by cloning restriction endonuclease sites into it. This phage contains single-stranded DNA, the form used in DNA sequencing by the Sanger method.

The most commonly used host microorganism for recombinant DNA studies has been *E. coli*. Its potential human pathogenicity raises some concerns for industrial production, and its inability to secrete enzymes makes product purification difficult for industrial usage. *Bacillus subtilis* is fairly well characterized genetically, but has not proven the perfect host for industrial applications. Among eukaryotic microbes, the yeast *Saccharomyces cerevisiae* is the best known in terms of its genetics. Therefore, its suitability as a host for eukaryotic gene expression has been investigated. Gene cloning in mammalian cells has used analogs of bacteriophage lambda; that is, animal viruses (such as SV40 and retroviruses) which integrate their genetic material into the host genome.

After the clone library has been established, the clone (bacterial colony) containing the recombinant DNA molecule of interest must be identified. If the foreign gene is expressed in the host bacterium and a protein is made, detection can be achieved by (1) a specific antibody reaction with the protein (see section 12.9 for details), (2) measuring the activity of the protein (if it is an enzyme, and this activity is not normally found in the host), or (3) complementing a mutation in the host bacterium, if the foreign gene is analogous to a host gene.

If the foreign gene is not expressed, then a radioactively labeled DNA fragment that is complementary to a portion of the foreign gene of interest is used as a **probe**. A replica copy of the bacterial colonies comprising the library is lysed to expose their DNA. The radioactive probe is added, and clones of interest are identified via hybridization of the probe to DNA in the colony.

In many cases, the purpose of cloning a gene is to obtain large quantities of its product. If the source of DNA is not an *E. coli* strain, it may not be expressed in an *E. coli* host. To remove this problem, **expression vectors** have been constructed in which the foreign gene is inserted in a configuration that puts it under regulatory controls recognized by the host bacterium. To maximize production of the foreign protein, the expression vector should replicate to a high copy number, the foreign gene should be linked to a strong promoter that has a high affinity for RNA polymerase and the mRNA should be efficiently translated. Another concern is that the foreign gene be linked to the regulatory elements such that the message is in the proper **reading frame** so that the correct codons   are   read.

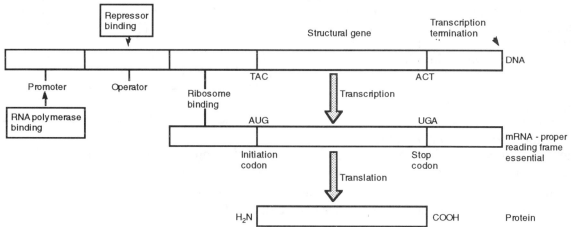

Expression of the cloned gene can be manipulated by placing it under the control of a **regulatory switch** such as the *lac*, *trp*, or lambda repressor systems. Production of the recombinant protein does not occur until the experimenter switches to the proper environmental conditions.

It is now possible to make short segments (30-35 bases) of DNA. The **synthetic DNA** is used in both research and genetic engineering. The most typical process used to make DNA is a solid-phase procedure where the first nucleotide is attached to a porous support and the segment grown out from this point.

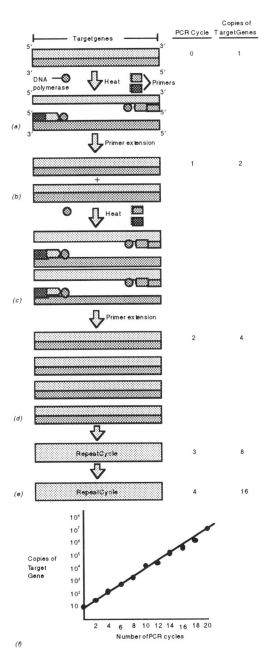

In several applications of genetic engineering, large amounts of specific DNA sequences are needed. The **polymerase chain reaction** can increase the amount of specific sequences by more than 10^6 fold. It entails multiple cycles of (1) heating DNA to produce single strands, (2) cooling and adding **primers** (short DNA sequence from the ends of the DNA to be amplified), and (3) primer extension catalyzed by thermostable *Taq* **polymerase**. The DNA polymerase must withstand the high temperatures required to produce single-stranded DNA in step 1. For this reason, the source of the enzyme is the thermophilic bacterium *Thermus aquaticus*. Each cycle doubles the content of the original target DNA. PCR has been combined with DNA fingerprinting procedures to allow identification of very small amounts of DNA.

If one wishes to clone and express genes from a eukaryotic organism in *E. coli*, the differences between eukaryotic and prokaryotic gene expression must be considered. Eukaryotes contain more DNA than bacteria, parts of it are non coding repetitive sequences, and the genome contains non coding introns that are excised by eukaryotes during RNA processing. For this last reason, one attempts to isolate the eukaryotic mRNA, and use **reverse transcriptase** to make a complementary DNA copy - - **cDNA**.

Identification of the specific mRNA involves hybridization to a DNA probe. The DNA probe is chemically synthesized, using existing information on the amino acid sequence of the protein made from the RNA. By consulting the genetic code, a nucleotide sequence can be constructed from the amino acid sequence. This probe is then used to identify specific mRNA, after the total mRNA population has been separated on the basis of size by agarose gel electrophoresis. An alternative method to enrich the mRNA of interest is to add antibodies that react with the protein produced from the **polysome**. These complexes contain a mRNA, ribosomes translating the message, and the polypeptide chain that is being synthesized. When the antibodies precipitate the specific protein, the mRNA is also precipitated.

93

Expression of eukaryotic genes in a prokaryote requires that the gene be inserted behind a strong promoter and a bacterial ribosome-binding site be present. One approach is to insert the eukaryotic DNA so that it is expressed as part of **fusion protein**. This combines some prokaryotic sequences of amino acid with the eukaryotic protein. The prokaryotic amino acid sequences are later removed and a pure eukaryotic protein results.

Genetic engineering has had a major impact on both basic and applied biology. One major impact in basic research is in the study of genetics. Bacterial geneticists can now specifically modify nucleotide sequences of interest by **site-directed mutagenesis**. Short oligonucleotides containing the desired mutation are chemically synthesized. This is paired in vitro with a single-stranded copy of the original gene. DNA polymerase will extend the oligonucleotide by complementary base pairing to make a complete double stranded copy of the gene. This is then inserted into a cloning vector, and introduced into a cell. In this way, specific amino acid substitutions can be made at specific sites in a protein, to test theories regarding the relationship between the protein's primary sequence and its function. Another approaches to genetic alteration is through **cassette mutagenesis**. Cassette mutagenesis is the removal of a large section -- cassettes -- of DNA. The original cassette is replaced by a second cassette that contains synthetic DNA in which one or more of the base pairs have been altered. It is also possible to replace the original cassette with a cassette for antibiotic resistance. This type of cassette mutagenesis is know as **gene disruption**. The resulting clone will be antibiotic resistant, but will have lost the gene function where the insertion took place.

The primary areas of commercial importance for recombinant DNA technology are in the production of (1) microbial products such as antibiotics, (2) vaccines that protect against viruses, (3) mammalian proteins, and (4) transgenic plants and animals. These techniques are also being used in the study of bacteria that degrade environmental pollutants and as a method to treat genetic disease.

Many mammalian proteins of medical importance are produced only in small quantities. Thus, it was not feasible to obtain them. By cloning these genes, bacteria could be used to obtain large amounts of these materials. Human **insulin**, human growth factor, parathyroid hormone, bone growth factor, and atriopeptin are examples of these types of proteins. The approach used to make human insulin illustrate that, beyond the problems in gene expression, further steps must be considered to produce a biologically functional product. Other mammalian products produced using genetically engineered microorganism include: **tissue plasminogen activator** (used to dissolve blood clots), **erythropoietin** (stimulates red blood cell production), any number of different **interferons** (anti-viral compounds), and **interleukin** (stimulates T lymphocytes) to name only a few compounds.

Existing virus vaccines are imperfect because they consist of killed virus. If all virus particles were not killed, an inoculation could cause infection. Our immune system responds only to the protein coat; thus by cloning the viral protein genes and producing only these proteins, the risk of viral infection from vaccination would be avoided. In some cases, an effective vaccine can be made by producing the viral proteins in bacteria. However, in other cases, this has been ineffective because the viral proteins are chemically modified (for example, by addition of sugars) after their production on ribosomes. This process does not occur in bacterial hosts.. For these cases, **subunit vaccines** are produced, in which the gene for coat protein of a pathogenic virus are cloned into a harmless virus, such as **vaccinia**.

Recombinant DNA techniques are being used to genetically alter plants. Foreign DNA can be introduced into plants by physical means such as electroporation or particle guns (see Section 7.6), or biologically by the bacterium *Agrobacterium tumefaciens*.. This plant pathogen induces tumor formation in plants by inserting part of its Ti plasmid DNA into plant chromosomes. The Ti plasmid can serve as a vector to insert foreign DNA into a plant cell. In some species, plant cells grown in culture can then be used to regenerate whole plants.

Transgenic animals can be produced by injecting recombinant DNA into fertilized eggs. New advances in biomedical research can be expected to arise from this technique. In addition, it may have commercial application, because human proteins produced by introducing cloned genes into animals would undergo the processing that occurs after translation, whereas these modifications do not occur when the genes are expressed in bacteria. **Gene therapy** is a therapeutic approach that allows an alteration of a dysfunctional gene directly in the patient. The first disease to be treated in this manner was for deficiencies of adenosine deaminase.

SELF TESTS

COMPLETION

1. The isolation and purification of genes is known as _____.

2. Isolated DNA molecules can be cut with _____ and put back together with _____.

3. Foreign DNA can be maintained within a host bacterium by inserting it into a _____.

4. A large group of bacterial cells, each containing different DNA fragments from another organism, is called a _____.

5. When plasmids are used as cloning vectors, recombinant DNA molecules are introduced into bacteria by _____.

6. When bacteriophage are used as cloning vectors, recombinant DNA molecules are introduced into bacteria by _____.

7. The amount of foreign DNA that can be cloned using lambda as a vector is limited by _____.

8. _____ has been the most commonly used host microorganism for recombinant DNA.

9. A clone library can be searched directly for a gene by the technique of _____.

10. When fusions are made between genes, it is important that the foreign gene be inserted in the correct _____.

11. The DNA from a eukaryotic cell will not lead to the correct production of the appropriate protein in a bacterium because it contains _____.

12. Genes can be cloned after isolating a messenger RNA by using the enzyme _____.

13. Site-directed mutagenesis involves the use of chemically synthesized _____ to replace specific portions of a gene.

14. The gene disruption technique relies on the insertion of a cassette containing antibiotic resistance and results in _____ of the gene function where the cassette is inserted.

15. cDNA is formed by _____ of mRNA.

16. Bacteriophage lambda, R-plasmids and certain other viruses can all serve as _____ for the insertion of DNA segments.

17. The polymerase chain reaction relies on primer extension by the enzyme _____ which is _____ stable.

18. Phagemids are vectors which contain both _____ and _____ origins of replication.

19. Cosmids are plasmid vectors which contain the _____ site from the lambda genome. This site is required for _____ DNA into the lambda virions.

20. The term clone can be used to indicate as few as _____ or as many as _____ microbial colonies.

KEY WORDS AND PHRASES

β-galactosidase (p 364)	autoradiography (p 371)
cDNA (p 381)	Charon phages (p 364)
codon usage (p 373)	colony hybridization (p 371)
colony hybridization (p 371)	cos sites (p 365)
cosmids (p 365)	diabetes (p 386)
DNA fingerprinting (p 378)	expression vector (p 372)
fusion protein (p 374)	gene disruption (p 383)
gene therapy (p 394)	*in vitro* recombination (p 361)
lambda P$_L$ promoter (p 373)	Northern blot (p 381)
oligonucleotide (p 376)	pBR322 (p 362)
phagemids (p 367)	polylinker (p 366)
polymerase chain reaction - PCR (p 376)	preproinsulin (p 386)
probe (p 360)	proinsulin (p 386)
promoter (p 373)	reading frame (p 373)
recombinant DNA (p 360)	replacement vectors (p 364)
restriction fragment length polymorphism (p 378)	reverse transcriptase (p 381)
reverse translation (p 381)	ribosome binding site (p 373)
secretion vector (p 366)	shuttle vector (p 366)
signal peptide (p 366)	site-directed mutagenesis (p 382)
synthetic DNA (p 375)	*tac* promoter (p 373)
transgenic animals (p 394)	transgenic plants (p 392)
vaccinia virus (p 389)	yeast artificial chromosome (p 367)

MULTIPLE CHOICE

1. Some of the following statements are incorrect with respect to the *in vitro* generation of recombinant DNA molecules:

    1. the DNA molecules must have a high degree of base sequence homology.

    2. the DNA molecules do not need to have a high degree of base sequence homology.

    3. the hybrid molecule must be able to replicate autonomously.

    4. the recombinant molecules are always introduced into the host cell by conjugation.

    5. a restriction enzyme is used to generate DNA molecules with single-stranded ends.

    (A) 1,2 (B) 2,3 (C) 3,4 (D) 4,5 (E) 1,4

2. Restriction endonucleases

    (A) act only on the DNA of the cell in which they are found

    (B) can be used to construct hybrid plasmid molecules

    (C) have very little specificity in the sites they attack

    (D) are primarily ribonucleases

    (E) only attack circular DNA

3. *Agrobacterium tumefaciens* contains a plasmid that

    (A) is a useful cloning vector for human DNA

    (B) induces a disease in plants

    (C) confers antibiotic resistance on a plant

    (D) is resistant to all restriction enzymes

    (E) is an expression vector for plant DNA

4. A recombinant DNA molecule in which the first third of the β-galactosidase gene was connected to a chemically synthesized DNA that had the code for human hemoglobin would lead to the production of

    (A) bacterial hemoglobin

    (B) a fusion protein

    (C) no protein. The message from this gene would not be translated.

    (D) shuttle protein

    (E) polysome

5. The most suitable cloning vectors have been

    (A) bacteriophage

    (B) plasmids

    (C) transposable elements

    (D) A and B above

    (E) all of the above

6. All of the following are desirable characteristics of a plasmid cloning vector EXCEPT:

    (A) small size

    (B) multiple cleavage sites for a restriction endonuclease

    (C) multiple antibiotic resistance markers

    (D) stable maintenance in a host bacterium

    (E) easy use in transformation

7. A specific clone in a library can be identified using

    (A) antibodies

    (B) nucleic acid probes

    (C) genetic complementation

    (D) enzyme assays

    (E) all of the above

8. A good expression vector should have all of the following EXCEPT:

    (A) a strong promoter

    (B) a Shine-Dalgarno sequence

    (C) a high copy number

    (D) high levels of repressor protein

9. The messenger RNA of eukaryotic cells can be distinguished from other cellular RNA because it contains

    (A) unique bases

    (B) poly A tails

    (C) unique restriction sites

    (D) ribosomes

    (E) secondary structure

10. Mutations in DNA can be detected by searching for

    (A) Shine-Dalgarno sequences

    (B) promoters

    (C) insertion elements

    (D) restriction fragment length polymorphisms

    (E) cDNA

11. The amount of a specific DNA sequence can be increased more than 10^6-fold by using which of the following chemical reactions?

  (A) restriction endonuclease reaction

  (B) ligation reaction

  (C) polymerase chain reaction

  (D) reverse translation

  (E) reverse transcriptase reaction

12. Scientists hope to produce transgenic plants which have improved resistance to:

  (A) herbicides

  (B) insects

  (C) microbial diseases

  (D) none of the above

  (E) all of the above

13. A newly isolated plasmid is characterized: the plasmid is 6,900 base pairs in size, has a single restriction site for *Hind*III, but multiple restriction sites for *Eco*RI, *Pst*I and *Sal*I. The plasmid lacks markers for antibiotic or any other resistance. How would you respond to the suggested use of the plasmid as a vector?

  (A) acceptable

  (B) acceptable if only *Hind*III is used to open the plasmid

  (C) acceptable if *Eco*RI, *Pst*I or *Sal*I are used to open the plasmid

  (D) acceptable if *Hind*III and *Sal*I are used to open the plasmid

  (E) not-acceptable, better vectors are available

14. Cosmids are best described as:

  (A) elements that eliminate problems associated with *E. coli* transformation

  (B) plasmid vectors containing foreign DNA and the cohesive ends of the lambda gene

  (C) plasmid cloning vectors constructed so they can contain large fragments of foreign DNA

  (D) packaged in vitro into lambda virions

  (E) all of the above

15. Which polymerase has made widespread use of PCR possible:

  (A) DNA polymerase I

  (B) *Thermus aquaticus* (*Taq*) polymerase

  (C) DNA polymerase III

  (D) none of the above

  (D) all of the above

FILL IN

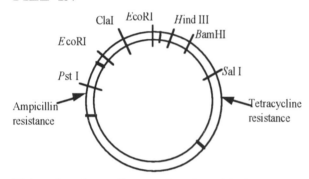

Using the above figure of a plasmid cloning vector, answer the following questions:

1. There are _____ restriction sites and _____ single insertion sites available.

2. Use of *Pst*I, followed by insertion of small piece of DNA would _____ ampicillin resistance.

3. In order to preserve both antibiotic resistance functions and plasmid integrity, a DNA insertion should be made at the _____ site.

4. In order to insert a large segment of DNA the plasmid could be opened at _____ site.

5. Following a DNA insertion at the *Sal*I site, clone selection could be based on _____ resistance.

6. Use of chloramphenicol in the growth media could allow up to _____ copies of the plasmid to be amplified.

7. An insertion of a segment of DNA that codes for enzymes that degrade 2,4,5 T (a herbicide) is made at the *Pst*I site. All of the resulting clones were resistant to tetracycline and 50% were resistant to ampicillin. Of these clones the portion you would expect to contain the 2,4,5 T gene would be _____?

DISCUSSION

1. What types of activities are included in the field of "genetic engineering"?

2. How might genetic engineering be used to produce safer vaccines?

3. Describe a mechanism by which desirable genes might be introduced into plants.

4. List three practical applications of genetic engineering.

5. What are the desirable properties of a cloning vector?

6. What modifications in phage lambda are made when it is used as a cloning vector?

7. What characteristics are desirable in a microbial host for cloned DNA?

8. What approaches would you use to find a foreign gene in a clone library when (a) the gene is expressed in the cloning host and (b) when it is not expressed?

9. What factors are important to the expression of a foreign gene in a bacterial cloning host?

10. What unique problems are encountered when one wants to express a eukaryotic gene in a bacterium?

ANSWERS

Completion

1. gene cloning; 2. restriction enzymes, DNA ligase; 3. cloning vector; 4. clone library; 5. transformation; 6. transduction; 7. size of the phage head; 8. Escherichia coli; 9. colony hybridization; 10. reading frame; 11. intervening sequences; 12. reverse transcriptase; 13. oligonucleotides; 14. disruption; 15. reverse transcriptase; 16. vectors; 17. Taq polymerase, heat; 18. plasmid, phage; 19. cos packaging; 20. one, unlimited.

Multiple choice

1. E; 2. B; 3. B; 4. B; 5. D; 6. B; 7. E; 8. D; 9. B; 10. D; 11. C; 12. E; 13. D (see Text section 8.2); 14. E; 15. B.

Fill in

1. 7, 5; 2. delete; 3. *Cla* I; 4. *Eco*R I; 5. ampicillin; 6. 3000; 7. zero.

Discussion

1. See introduction to Chapter 8, Table 10.1, text Section 10.12 - 8.15.

2. See text Section 10.12.

3. See text Section 10.12.

4. See text Section 10.12-10.13

5. See text Section 10.2-10.4.

6. See text Section 10.3.

7. See text Section 10.5.

8. See text Section 10.6.

9. See text Section 10.7.

10. See text Section 10.10.

Microbial Growth Control

OVERVIEW

Chapter 11 (pages 399-431) concerns the control of microbial growth. In previous chapters, the factors that promote growth have been discussed; this chapter deals with inhibition and prevention of microbial growth, including **disinfection**, **sterilization** and **chemotherapy**.

CHAPTER NOTES

Control of microbial growth can be by inhibition of growth, killing the microorganisms or removing them from an environment. Antimicrobial agents can be divided into agents that kill microorganisms (**bactericidal**) and agents that inhibit growth (**bacteriostatic**) of the microbes. **Sterilization** is the process of killing or removing all living organisms and viruses an environment. Sterilization, may be accomplished by several physical methods, including **heat**, **filtration**, and **radiation**. Of these methods, heat is by far the most common. As the temperature increases beyond the maximum temperature for growth of a microorganism, lethal effects occur. The rate of death is a function of both the temperature and

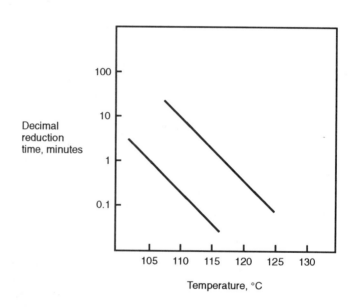

the time of exposure. Microbial death is an exponential function (death is linear when plotted on a log scale. The **decimal reduction time** is the time required to reduce the population by a factor of 10. In the figure, the upper line represents a heat resistant microorganism.

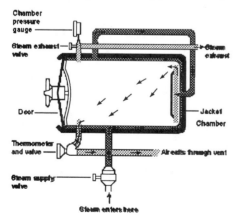

The autoclave typically uses an operating temperature of 121°C. At this temperature, bacterial endospores have a decimal reduction time of 4 - 5 min, whereas vegetative cells have decimal reduction times in the range of 0.1 to 0.5 min. Because endospores are not killed boiling temperatures (100°C), the autoclave uses steam under pressure, typically 15 lb/in. Generally, the time after the autoclave reaches pressure is 10-15 min, however, bulky objects or large volumes of liquid require more time.

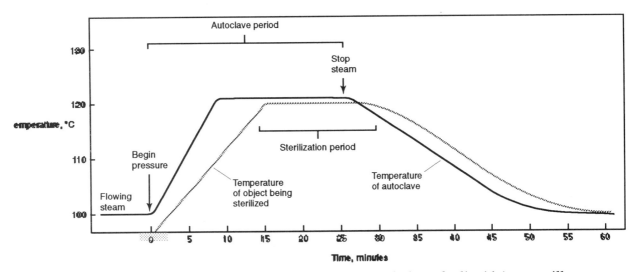

Pasteurization is a process that reduces the microbial population of a liquid (e. g., milk, wine, and fruit juices). The process is named for **Louis Pasteur**, whom first used the methodology for making wine (see the box). During **bulk pasteurization**, the liquid is held in vats at 63 to 65°C for 30 min. In **flash pasteurization**, which gives more satisfactory results and can be adapted to **continuous-flow**, the liquid is heated to 71°C for 15 sec, then rapidly cooled. Pasteurization extends the shelf life of a product and reduces the level of pathogens in the product.

Electromagnetic irradiation is another effective way to sterilize or reduce **microbial burden** of almost any substance. **Microwaves**, **ultraviolet (UV) radiation**, **X-rays**, **gamma rays** and **electrons** are used although each type of irradiation has a specific mechanism. UV irradiation, which does not penetrate solid, opaque or light absorbing materials, is useful for disinfecting surfaces, air and liquids that do not absorb the UV waves. Gamma and X-rays, which are more penetrating, are more difficult and expensive to use but are finding application in food preservation and other industrial processes. Irradiation has become a useful alternative to **ethylene oxide** in preparation of surgical supplies.

Microorganisms may be removed from liquid media by **filter sterilization**. **Depth filters**, which are made from paper, asbestos or glass fibers, consist of a random array of overlapping fibers. Depth filters trap particles in the torturous paths created throughout the depth of the structure. **Membrane filters** are composed of cellulose acetate or cellulose nitrate and are made in such a way that the filter contains a large number of tiny holes. Thus, the microorganisms are trapped on the surface of the filter. Finally **nucleation track filters** are created by treating very thin polycarbonate films with nuclear radiation then etching the film with a chemical. The sizes of the holes can be precisely controlled. Sterilization by filtration may preserve biologically important molecules that are inactivated by heat.

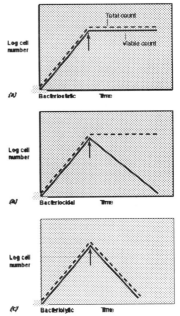

Antimicrobial agents are chemicals that kill or inhibit microorganisms. **Cidal agents** kill microbes, thus the terms **bactericidal, fungicidal, and viricidal.** **Static agents (bacteriostatic, fungistatic and viristatic)** inhibit growth. Antimicrobial agents act in a number of ways as shown in the figure to the left.

103

The **minimum inhibitory concentration (MIC)** is the amount of the agent required to inhibit growth of the test organism. MIC's are often determined by the **tube dilution technique**, which uses various concentrations of antimicrobial agents in medium containing a constant **inoculum** of microorganisms. The method varies with the culture medium, inoculum size, incubation time, and nature of the test organism. Antimicrobial action may also be studied by the **agar diffusion method**. A Petri plate containing an agar medium is evenly inoculated with the test organism. Known amounts of the antimicrobial agent are added to filter paper disks that are placed on the surface of the agar. During incubation, the agent diffuses from the disk and results in a **zone of clearing** around the disk that can be measured -- larger zones of clearing imply more effective agents. **Disinfectants** are chemical antimicrobial agents, frequently termed **germicides,** that are used on inanimate objects, whereas **antiseptics** are agents used on living tissue.

Control of microbes is important in the food industry because some microbes spoil food or their growth in food produces toxins. The susceptibility of a product to food spoilage is a consequence of its suitability as a growth medium. Therefore, foods with low water activity are less susceptible to **microbial spoilage**. Food can be preserved by (1) lowering the storage temperature, (2) lowering the pH, or (3) adding sugar or salt to decrease water activity. Canning is a process in which a food is sealed and heated so as to kill all living organisms or at least to ensure that there will be no growth of residual organisms in the can. Thus, Canning is a type of heat sterilization.

Chemotherapeutic agents are agents used for control of disease. **Growth factor analogs**, are substances that are related to growth factors and block utilization of the growth factor in the infectious agent. The first of these to be discovered was the **sulfa drugs**. One of these, **sulfanilmide** acts as an analog of *p*-aminobenzoic acid and blocks synthesis of folic acid.

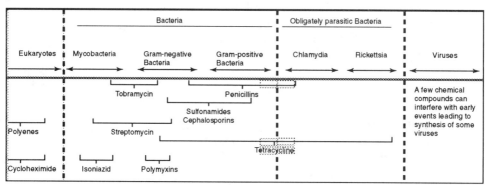

Antibiotics are microbial or synthetic substances that inhibit other microbial species. Only a few of these are practical in treating infectious diseases, because they may cause toxic effects upon animals or are ineffective in the body. The effectiveness of an antibiotic may be enhanced by chemical modification after its microbial production. Targets of action include the bacterial cell wall, membrane, and ribosomes. Bacteria may be innately resistant to an antibiotic, or may acquire resistance. Resistance can result from (1) the absence of the target of action, (2) drug impermeability, or (3) production of an inactivating enzyme. Antibiotics are a unique class of chemotherapeutic agents because they are natural products rather than synthetic chemicals, such as the sulfa drugs.

104

SELF TESTS

COMPLETION

1. Agents that destroy or kill bacteria are called _____ agents.

2. Uncontrolled microbial growth causing cell destruction in a host is called _____.

3. When microbes are killed by irradiation, the amount of radiation required to accomplish a 10-fold reduction in the population is called the _____.

4. The use of heat, such as heating a liquid to 63°C for 30 min followed by cooling, is referred to as _____.

5. The _____ is a sealed device that allows the entrance of steam under pressure and uses moist heat to kill all microorganisms, including endospores.

6. An agent that kills or inhibits microorganisms is called a _____.

7. An agent that inhibits, but does not kill a bacterium, is called a _____ agent.

8. Chemicals that are used to control microbes on inanimate objects are called _____.

9. Any change in the visual appearance, smell or taste of a food product that makes in unacceptable to the consumer is called _____.

10. Agents used to internally to control disease are called _____.

11. _____ are substances made by microorganisms to kill or inhibit other microorganisms.

12. The β-lactam antibiotics include the _____ and the _____.

KEY WORDS AND PHRASES

β-lactam antibiotic (p 417)	β-lactamase (p 427)
agar diffusion method (p 409)	aminoglycoside (p 420)
antibiotic (p 416)	antimicrobial agent (p 407)
autoclave (p 402)	bactericidal (p 400)
bacteriolytic (p 408)	bacteriostatic (p 400)
canning (p 413)	cephalosporin (p 420)
chemotherapeutic agent (p 414)	cidal agent (p 407)
decimal reduction time (p 401)	depth filter (p 406
disinfectants (p 410)	drug resistance (p 426)
erythromycin (p 421)	filter sterilization (p 406
Food and Drug Administration (p 403)	food preservation (p 412)
food spoilage (p 411)	fungicidal (p 407)
germicide (p 410)	growth factor analog (p 414)
inhibition (p 400)	ionizing radiation (p 403)
macrolide (p 421)	membrane filter (p 406)
nucleation track filter (p 406)	pasteurization (p 402)

penicillin (p 417)	penicillin binding protein (p 419)
penicillin G (p 417)	selective toxicity (p 407)
static agent (p 407)	sterilization (p 400)
streptomycin (p410)	sulfa drug (p 414)
sulfanimide (p 414)	tetracycline (p 421)
thermal death time (p 401)	tube dilution technique (p 409)
UV radiation (p 402)	viricidal (p 407)

MULTIPLE CHOICE

1. A chemical that causes bacterial death would be considered

    (A) bactericidal

    (B) bacteriostatic

    (C) bacteriolytic

    (D) depending on the chemical, A or B may be correct

    (E) depending on the chemical, A or C may be correct

2. The time required to reduce a microbial population by 10-fold is referred to as

    (A) the decimal reduction value

    (B) the decimal reduction time

    (C) the thermal death time

    (D) the log time

    (E) the thermal reduction value

3. During sterilization using an autoclave, one is most concerned with survival of

    (A) *Mycobacterium tuberculosis*

    (B) mold spores

    (C) vegetative cells

    (D) bacterial endospores

    (E) viruses

4. The typical temperature for an autoclave during the sterilization process is

    (A) 121°C   (D) 200°C

    (B) 100°C   (E) 63°C

    (C) 73°C

5. Pasteurization might be used on

    (A) wine   (C) beer

    (B) vinegar   (D) milk

    (E) All of these

6 UV radiation might be an effective way to disinfect which of the following

(A) contaminated hamburger

(C) distilled water flowing through a pipe

(B) a paper-wrapped glass tube

(D) a sealed box of soil

(E) All of the above

7. X-ray irradiation might be an effective way to disinfect which of the following

(A) contaminated hamburger

(C) distilled water flowing through a pipe

(B) a paper-wrapped glass tube

(D) a sealed box of soil

(E) All of the above

8. The _____ filter (for filter sterilization) requires the least advanced technology during manufacture.

(A) depth filter

(B) the membrane filter

(C) the nucleation track filter

(D) All three of the above require similar technologic development.

9. Which of the foods listed would be most likely to spoil as a result of an enteric bacterium?

(A) A meat with near neutral pH.

(B) A fruit in a high sugar syrup

(C) a vegetable in a high salt, acid brine.

(D) All of the above are likely to suffer spoilage as the result of bacterial growth.

(E) There is no basis on which to make this judgment.

10. Which temperature is most common for storage of frozen food?

(A) 0°C

(D) -20°C

(B) 15°C

(E) -70°C

(C) 4 °C

11. Which of the following chemicals is not commonly included in a food product as a preservative?

(A) sodium benzoate

(D) sodium nitrite

(B) ethylene oxide

(E) All of the above are commonly used

(C) sorbic acid

12. _____ have (has) been successful targets for growth factor analogs.

(A) Folic acid

(C) Amino acid

(B) Vitamins

(D) Purines and pyrimidines

(E) All of the above are known targets for growth factor analogs

13. _____ is a drug that inhibits bacterial cell wall synthesis.

    (A) penicillin

    (B) rifampicin

    (C) tetracycline

    (D) sulfa drugs

    (E) All of the above

14. _____ is a drug that inhibits bacterial protein synthesis.

    (A) penicillin

    (B) rifampicin

    (C) tetracycline

    (D) sulfa drugs

    (E) All of the above

15. _____ is a drug that inhibits bacterial folic acid synthesis.

    (A) penicillin

    (B) rifampicin

    (C) tetracycline

    (D) sulfa drugs

    (E) All of the above

MATCHING

Match the drug with its associated property

Drug	Associated property
1 tetracycline	A) blocks protein synthesis at 30S ribosome
2. penicillin	B) blocks protein synthesis at 30S ribosome
3. sulfanilamide	C) an antiviral drug
4. streptomycin	D) β-lactamase sensitive
5. AZT	E) an analog of p-aminobenzoic acid

DISCUSSION

1. Describe how an autoclave works from a mechanical point of view.

2. If you were working in a developing country and wanted to develop a filter sterilization using minimal resources, which kind of filter system would you choose? Why?

3. Compare and contrast chemical agents that are used for disinfectants and antiseptics. Are the same chemicals used for both?

4. Why might an mold, such as *Penicillium* produce an antibiotic?

5. In general, why has the development of antiviral drugs been slow, compared to antibacterial agents?

6. What factors have contributed to the development of antibiotic resistance during recent years?

ANSWERS

Completion

1. bactericidal; 2. infectious disease; 3. decimal reduction value; 4. pasteurization; 5. autoclave; 6. antimicrobial agent; 7. bacteriostatic agent; 8. disinfectants; 9. food spoilage; 10. chemotherapeutic agents; 11. Antibiotics; 12. penicillins, cephalosporins.

Multiple Choice

1. E; 2. B; 3. D; 4. A; 5. E; 6. C; 7, E; 8. A; 9. A; 10, D; 11. B; 12. E; 13. A; 14. C; 15. D

Matching

1. B; 2. D; 3. E; 4. A; 5. C.

Discussion

1. See text section 11.1

2. See text section 11.3

3. See text section 11.5

4. See text sections 11.8 and 11.9.

5. See text sections 11.8 and 11.11.

6. See text section 11.13

Chapter 12
INDUSTRIAL MICROBIOLOGY

OVERVIEW

Chapter 12 (pages 432-474) concerns itself with microorganisms that have been used in commercial processes such as beverage production for thousands of years. Over the past hundred years, pure cultures of microbes have been used to produce products that have commercial value.

CHAPTER NOTES

The term **fermentor** when used in the industrial microbiology context, includes any large scale microbial process carried out under aerobic or anaerobic conditions. Most recently, genetically engineered bacteria have been grown on an industrial scale to produce substances they do not normally make. It should also be noted that microbiologists use the term **fermentation** in two different contexts. In the context of metabolism, fermentation refers to growth in the absence of an external electron receptor whereas in the context of industrial microbiology, the term refers to the growth of large quantities of cells.

Industrial microorganisms are initially selected from natural samples, or taken from a **culture collection** because they have been shown to produce a desired product. However, the strain is then modified to improve the product yield. This entails rounds of mutation, with careful selection for the rare clones that produce more or improved products. The selected strain is unlikely to survive well in nature, because the selection process has altered the regulatory controls in the cell to create metabolic imbalances. Other desirable characteristics are (1) rapid growth, (2) genetic stability, (3) non-toxicity to humans, and (4) large cell size, for easy removal from the culture fluid. There are several important culture collections that maintain repositories of important microorganisms. In the United States, the **American Type Culture Collection (ATCC)** is probably the best known, however, the Northern Regional Research Laboratory (NRRL) is also well known. Table 12.1 provides a list of other important culture collections.

Industrial microbiologists may aim to produce (1) microbial biomass itself, (2) specific enzymes, or (3) metabolites. Enzymes that degrade polymers are especially important. The metabolites may be major metabolic products of catabolism, or compounds normally produced in trace amounts by natural isolates. Industrial microbiology is an issue of scale. Although industrial microbiologists culture organisms in many of the same ways as other microbiologists, the goal is often to produce very large quantities, sometimes measured in millions of liters at one time.

The pharmaceutical industry is an important user of microbes. For many years, **antibiotics** and **steroid hormones** have been produced by microbes. Genetic engineering has made it possible for bacteria to produce a wide variety of mammalian substances that are medically important.

In agriculture, bacteria of the genus *Rhizobium* are added to legume seeds where, following nodule formation on the host plant, they trap or fix atmospheric nitrogen and reduce the need for ammonia fertilizer.

Several **specialty chemicals** are most economically produced by microbes. These include some amino acids and vitamins. Industrial solvents like alcohols and acetone can be produced in microbial fermentations. However, at present it is cheaper to make them chemically from petroleum.

Many of the most important industrial metabolites are **secondary metabolites**, produced in the stationary phase of the culture after microbial biomass production has peaked. These compounds are not essential for growth of the microbe. Their synthesis is usually tightly regulated by the cell. Therefore to obtain high yields, environmental conditions that elicit regulatory mechanisms such as repression and feedback inhibition must be avoided. In addition, mutant strains that **overproduce** the compound are selected. In secondary metabolism two phases are apparent: **trophophase** and **idiophase**. Trophophase is the growth phase of the culture; idiophase is the time when the secondary metabolites are formed. The success of the idiophase is dependent on the trophophase.

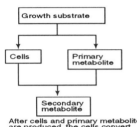

Cells and metabolite are produced more or less simultaneously

After a suitable microorganism has been identified from laboratory studies for an industrial process, there are still a number of **scale-up** problems to solve. These include provision of adequate aeration and mixing throughout the large fermentor. The difficulties involve the enormous volume of the vessel, areas where mixing is less efficient, and the high biomass content of the fermentor. High biomass is desirable to increase the product formed, but it creates an enormous demand for oxygen. Furthermore, a strain that worked well on a small scale may not be as efficient under the different conditions experienced in the large fermentor.

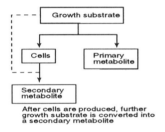

After cells and primary metabolite are produced, the cells convert the primary metabolite to a secondary metabolite

After cells are produced, further growth substrate is converted into a secondary metabolite

For industrial processes, fermentors of capacity up to 400,000 liters are used. The process may be aerobic or anaerobic. In general, aerobic processes are more difficult to run because it is difficult to adequately aerate a large vessel that contains high biomass concentrations. The fermentors are constructed of stainless steel, and have an external jacket by which it can be sterilized initially and cooled during the fermentation. **Spargers** and **impellers** in the vessel are used for aeration and stirring of the contents. The vessel may contain various devices for monitoring the environmental conditions within the culture, so that these factors can be controlled to obtain high product yields.

The organisms used in industrial processes must be carefully preserved so that their carefully selected attributes do not change because of mutation. This may involve frozen storage in liquid nitrogen or lyophilization. The inoculum for the fermentor must be built up from a **working strain**. Since inocula should be 5-10% of the culture volume, the inoculum for a production fermentor may be 10,000 liters.

Antibiotics are among the most important compounds produced by industrial microbes. The most useful ones are secondary metabolites produced by filamentous fungi, and bacteria classified as **actinomycetes**. New antibiotics are discovered by **screening** microbes isolated from natural samples for the production of chemicals that inhibit specific test bacteria. The test bacteria are related to bacterial pathogens. Most of the positive results are likely to be currently known antibiotics, but new ones are still discovered. The new substance is tested for toxicity and effectiveness in infected animals before commercial production is contemplated. Most new substances are likely to be either toxic to the animal, or relatively ineffective in killing pathogens in the body. Thus, an enormous amount of effort must be expended to find the rare substance that is novel and effective.

If a compound passes these tests, then commercial production can be contemplated. A high-yielding strain must be selected, and a purification procedure established. Antibiotics are synthesized by complex biochemical pathways. A great deal of research goes into understanding these pathways, to indicate ways of selecting higher yielding strains.

β-lactam antibiotics include the penicillins and cephalosporins. They are especially useful because their target of action is a compound not contained in any eukaryote -- the peptidoglycan of the cell wall. They inhibit its synthesis by inhibiting the transpeptidation reaction.

Streptomycin, an **aminoglycoside** antibiotic, illustrates that although antibiotics are designed to be antibacterial, they may cause side effects in animals. Such compounds are kept in reserve, for use in life-threatening situations where other antibiotics have failed.

Tetracyclines are **broad-spectrum** antibiotics effective against both Gram positive and Gram negative species. These and the β-lactam antibiotics are most useful medically. They have been used in non medical applications such as animal supplements, but these uses promote the spread of plasmid-encoded antibiotic resistance genes, and make antibiotic therapy potentially less successful in controlling serious diseases.

Vitamins and amino acids are used as supplements in human and animal feed. Some of these are produced most economically by bacteria, if high-yielding, overproducing strains can be developed. In general, this involves inactivating the regulatory mechanisms that keep biosynthesis of these substances in line with that of other cellular building blocks. One trick for doing so is to find mutants that grow in the presence of chemical analogs of the amino acid. These often lack feedback control of enzyme activity.

In some industrial processes, microbes are only used to carry out a specific biochemical reaction; the remainder of product formation is accomplished by strictly chemical means. An example of this type of **bioconversion** is steps in the production of steroid hormones.

The largest use of microbial enzymes has involved extracellular hydrolytic enzymes that digest insoluble materials. They have been used in the laundry industry; one problem is that some people are allergic to these proteins. A suite of three enzymes is used to produce a sweet material from starch. **High-fructose corn syrup** is produced by hydrolyzing starch to glucose, and **isomerizing** glucose to a sweeter molecule, fructose. Another important microbial enzyme is microbial rennin that is widely used in the cheese production.

Enzymes can be immobilized so that they are easier to recover and reuse, or because it stabilizes them. They can be cross-linked to each other, bonded to an inert material, or encapsulated within a membrane.

Vinegar can be produced by the **acetic acid bacteria** from an alcoholic fluid such as wine or cider, if oxygen is provided. There are several industrial methods for bringing these reactants together.

Citric acid is produced by the fungus *Aspergillus niger*. Industrial fungal fermentations may occur on the surface of a medium or submerged in the liquid. For **surface processes**, the medium can be either solid or liquid.

Yeast cells are the most extensively used microbe in industry. They are grown for use as **baker's yeast** in bread dough production. Under anaerobic conditions, they ferment sugars to alcohol, and form the basis for the wine and beer industries. Most industrial processes use various strains of one yeast species, *Saccharomyces cerevisiae*.

Wine is the fermented juice of fruits, usually grapes. **Must** is produced by crushing the grapes, and the wine yeast is added. Wild yeasts may be killed before inoculation by treatment with sulfur dioxide. After approximately two weeks, the fermentation has produced 10-12% alcohol. **Racking** the wine clarifies it and promotes flavor development.

Beer is produced by fermenting malted grains. The grains are prepared by **mashing**, in which they are cooked and steeped in water. Enzymes in malt digest the starch to sugars that can be fermented by yeast. The resulting **wort** is boiled to sterilize the solution after the flavoring ingredient **hops** is added. Brewery yeasts are added, and the fermentation proceeds for 1-2 weeks, depending upon the type of beer produced. Lager beer is then stored in the cold for several weeks.

Microbes have potential as food supplements because many of them contain 50% or more protein. In most diets, protein is in shortest supply. However, there can be problems with toxicity and digestibility that limit the amount of **single-cell protein** in the human diet. Microorganisms are important in wastewater treatment for two reasons. First, one of the goals of treatment is to destroy all pathogenic microbes that are in the sewage. Second, microbial activity is used to oxidize the organic matter in wastewater to methane or carbon dioxide gas.

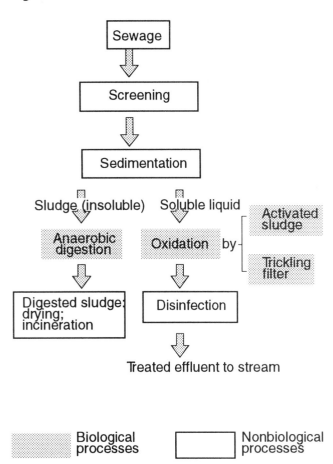

There can be three stages of sewage treatment. **Primary treatment** physically removes particulate material with screens and in a settling pond. **Secondary treatment** uses microorganisms to reduce the level of organic matter; the amount of residual organic matter is quantified as the **biochemical oxygen demand**. **Tertiary treatment** includes processes that remove inorganic nutrients, such as phosphate and nitrate, from the wastewater. Most sewage plants use only primary and secondary treatment of wastewater.

Secondary wastewater treatment may involve either aerobic or anaerobic processes. Under anaerobic conditions, the microbial reactions are similar to what was described in Sections 17.11 and 17.12. Macromolecules are hydrolyzed to soluble monomers; these are fermented by a series of bacteria to acetate, H_2, and CO_2. These substrates are converted to methane gas by methanogens.

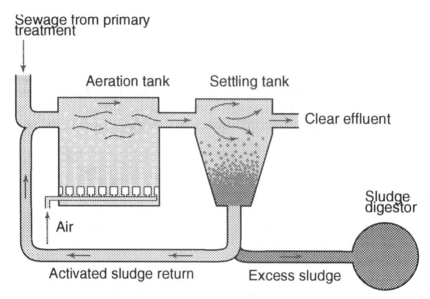

Sewage from primary treatment

Aeration tank

Settling tank

Clear effluent

Sludge digestor

Air

Activated sludge return

Excess sludge

Activated sludge treatment

Aerobic secondary treatment takes place in either the **trickling filter** or the **activated sludge** process. In the latter, flocs of microbes form in which organic matter is either oxidized or adsorbed. The flocs settle in a holding tank, and are added to an anaerobic sludge digestor, where the remainder of the organic material is converted to gaseous products.

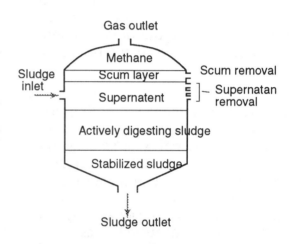

Gas outlet

Methane

Scum layer — Scum removal

Sludge inlet

Supernatent — Supernatan removal

Actively digesting sludge

Stabilized sludge

Sludge outlet

SELF TESTS

COMPLETION

1. The yield of an industrial fermentation can be increased by improving the cultural conditions or by _____.

2. The recovery and stability of industrial enzymes is enhanced if they are _____.

3. Citric acid is produced commercially by the fungus _____.

4. Vinegar is produced from _____ by the metabolism of _____.

5. If starch is the starting material for an alcoholic fermentation, the starch must first be _____ to _____ before it is fermented.

6. Bread dough rises because bakers' yeast produces _____ from sugar by fermentation.

7. Sulfur dioxide is added to wine to _____ wild strains of yeasts.

8. Many important microbial products are _____ metabolites, formed during the stationary phase of growth.

9. In industrial scale fermentors, the environmental factor which is most difficult to control is _____.

10. The initial step in finding a new antibiotic is _____.

11. _____ penicillins are produced by a combination of fermentation with chemical modification.

12. The _____ are a medically useful group of broad-spectrum antibiotics.

13. Microbial production of amino acids is economical when only the _____-form of the compound is desired.

14. The metabolic activity of yeast is used in bakery products to generate _____.

15. The aging of distilled alcoholic beverages in wood casks removes _____.

16. When making fortified wine, _____ is added after the fermentation.

17. The difference between red and white wine is that with red wine production the _____ is not removed during the fermentation process.

19. Much of the carbon found in solid material in sewage wastes ultimately ends up as _____ gas.

KEY WORDS AND PHRASES

β-lactam antibiotic (p 444)	acetic acid bacteria (p 456)
activated sludge (p 469	*Agaricus bisporus* (p 466)
ale (p 463)	anaerobic sludge digestor (p 470)
aspartame (p 450)	bioconversion (p 452)
biosynthetic penicillin (p 444)	bottom yeasts (p 462)
compressed yeast cake (p 459)	cooling jacket (p 438)
corn steep liquor (p 435)	cross-streak method (p 443)
culture collection (p 434)	distilled vinegar (p 456)
dry wine (p 460)	fortified wine (p 460)
gray water (467)	hops (p 462)
idiophase (p 437)	immobilized enzyme (p 454)
impeller (p 438)	industrial fermentation (p 433)
lager beer (p 463)	microbial products (p 433)
modified organisms (p 434)	mushrooms (p 466)
must (p 460)	mutant (p 451)
natural penicillin (p 444)	orleans method (p 456)
pilot plant (p 441)	primary metabolite (p 436)

primary treatment (p 468)	protein supplement (p 466)
racking (p 461)	riboflavin (p 450)
rum (p 463)	*Saccharomyces carlsbergensis* (p 463)
Saccharomyces cerevisiae (p 458)	secondary metabolite (p 436)
secondary treatment (p 469)	semisynthetic penicillin (p 444)
sewage (p 467)	shiitake (p 466)
single-cell protein (p 435)	sparger (p 438)
sparkling wine (p 460)	tertiary treatment (p 469)
top yeasts (p 462)	trickling filter (p 469)
trophophase (p 437)	vodka (p 463)
wine bouquet (p 460)	wort (462)

MULTIPLE CHOICE

1. The yield of a microbial product may be increased if the cell has multiple copies of the appropriate genes. This is called:

   (A) induction
   (B) gene fusion
   (C) gene amplification
   (D) gene vectoring
   (E) gene therapy

2. Mutations that increase the yield of a microbial product are most likely to occur in:

   (A) biosynthetic genes
   (B) catabolic genes
   (C) regulatory genes
   (D) plasmid-encoded genes

3. All of the following are desirable characteristics for an industrial microorganism EXCEPT:

   (A) large size
   (B) fast growth rate
   (C) genetically stable
   (D) amenable to genetic manipulation
   (E) can grow in humans

4. Commercially important antibiotics are produced by:

   (A) acetic acid bacteria
   (B) actinomycetes
   (C) yeast
   (D) lactic acid bacteria
   (E) fermentative bacteria

5. The fungus *Penicillium* produces which of the following antibiotics?

   (A) ampicillin
   (B) methicillin
   (C) oxacillin
   (D) penicillin-G
   (E) all of the above

116

6. Which of the following are β-lactam antibiotics?

    (A) penicillin-G   (D) cephalosporin-C

    (B) ampicillin   (E)   all of the above

    (C) methicillin

7. β-lactamase is a(n)

    (A) polysaccharide   (D) antibiotic

    (B) enzyme   (E) coenzyme

    (C) cell structure

8. In the production of high-fructose corn syrup, glucose is converted to a sweeter molecule by

    (A) amylase   (D) fructose aldolase

    (B) glucoamylase   (E) aspartase

    (C) glucose isomerase

9. In industrial fermentations, alcohol is produced from sugars by:

    (A) *Penicillium*   (D) *Saccharomyces*

    (B) *Cephalosporium*   (E) *Aspergillus*

    (C) *Acetobacter*

10. Both beer and wine are produced by fermentation of grains or grapes by:

    (A) acetic acid bacteria

    (B) *Aspergillus niger*

    (C) *Saccharomyces cerevisiae*

    (D) *Agrobacterium tumefaciens*

    (E) lactic acid bacteria

11. The red color in some wines is due to

    (A) the yeast which is used   (D) the grape skins

    (B) artificial coloring   (E) sulfur dioxide treatment

    (C) the temperature of the fermentation

12. The alcohol content of unfortified wines is:

    (A) 3% or less   (D) 30-35%

    (B) 8-14%   (E) 40-50%

    (C) 20-25%

13. The production of secondary metabolites occurs at or near the onset of which microbial growth phase:

    (A) lag phase   (D) death phase

    (B) stationary phase   (E) all of the above

    (C) exponential phase

14. The production of high-fructose corn syrup, used in soft drinks, is a microbial process. How do microorganisms play a role in the formation of the product?

(A) Microbial populations growing in fermentors, convert corn to the product

(B) Three microbial enzymes (α-amylase, glucoamylase and glucose isomerase) are used to convert corn starch to fructose.

(C) Fructose is formed by isolates of *Saccharomyces cerevisiae* growing on corn

(D) Microorganisms are not used in the production of soft drinks

(E) Microbes produce rennin which is then used as coloring agent for soft drinks

15. Yeast has been used extensively to make all of the following EXCEPT:

(A) food for animals and people

(B) B and D vitamins

(C) beer, wine, and distilled beverages

(D) ATP and invertase

(E) penicillin and streptomycin

16. Sewage treatment involves _____ biological processes.

(A) aerobic

(B) anaerobic

(C) aerobic and anaerobic

MATCHING

I. Match the antibiotic with the producing organism

Antibiotic	**Microorganism**
1. tetracycline	(A) *Streptomyces erythreus*
2. kanamycin	(B) *Streptomyces kanamyceticus*
3. penicillin	(C) *Streptomyces griseus*
4. polymyxin B	(D) *Penicillium chrysogenum*
5. cycloheximide	(E) *Streptomyces rimosus*
6. erythromycin	(F) *Bacillus polymyxa*

II. Match the microbial enzyme with its action.

Enzyme	**Action**
1. α-amylase	(A) converts glucose to fructose
2. invertase	(B) sucrose digesting
3. protease	(C) production of glucose monomers
4. rennin	(D) starch polysaccharide chain shortening
5. glucose isomerase	(E) coagulation of milk
6. glucoamylase	(F) protein breakdown

118

DISCUSSION

1. What techniques are used to improve the yield of product formed by an industrial microorganism?

2. What components or products of microorganisms have commercial value?

3. What factors keep the amount of secondary metabolites produced by normal cultures at low levels? How does the industrial microbiologist circumvent these factors to generate a high-yielding fermentation?

4. In what ways does an industrial-scale fermentor differ from the culture vessels used in the laboratory? What environmental factors must be controlled in the fermentors?

5. What specific problems are encountered in trying to scale up a process from the laboratory to an industrial fermentor?

6. Describe the steps that are involved in finding a new, medically useful antibiotic.

7. What regulatory mechanisms must be bypassed to achieve overproduction of an amino acid in an industrial process?

8. In what ways are enzymes derived from microbes used in commercial processes?

9. What steps, other than the fermentation of sugar to alcohol, are associated with the production of (a) wine, (b) beer, and (c) distilled liquor

10. What advantages and disadvantages are there to the use of microbes as a food supplement for humans?

11. Compare the relative advantages and disadvantages of aerobic and anaerobic fermentors for the production of a given product.

12. In scaling up a microbial process from a small (300 L) aerobic pilot fermentor to a full scale fermentor (200,000 L) the production of a number of unwanted secondary materials results. You as the industrial microbiologist are called on to **fix** the problem. Describe where you would begin your assessment.

13. Compare and contrast a trickling filter with an activated sludge system.

ANSWERS

Completion

1. selecting mutants; 2. immobilized; 3. *Aspergillus niger*; 4. ethanol, acetic acid bacteria; 5. hydrolyzed, glucose; 6. carbon dioxide; 7. kill, recalcitrant; 8. secondary; 9. aeration; 10. screening natural isolates; 11. semisynthetic; 12. tetracyclines; 13. L; 14. carbon dioxide; 15. fusel oils; 16. alcohol; 17. grape skins and stems; 18. methane.

Multiple choice

1. C; 2. C; 3. E; 4. B; 5. D; 6. E; 7. B; 8. C; 9. D; 10. C; 11. D; 12. B; 13. B;

14. B; 15. E; 16. C.

Matching

I. 1. E; 2. B; 3. D; 4. F; 5. C; 6. A.

II. 1. D; 2. B; 3. F. 4. E; 5. A. 6. C.

Discussion

1. See text Chapter introduction and Section 12.1

2. See text Section 12.2 and 12.3

3. See text Section 10.2
4. See text Section 12.3
5. See text Sections 12.3 and 12.4
6. See text Sections 12.5, and 12.6
7. See text Section 12.7
8. See text Section 12.9
9. See text Section 12.13
10. See text Section 12.14
11. See text Section 12.3
12. See text Section 12.4
13. See text Section 12.15

Chapter 13
METABOLIC DIVERSITY AMONG THE MICROORGANISMS

OVERVIEW

Chapter 13 (pages 575-622) concerns the diversity in metabolic abilities that is encompassed by the world of microorganisms. A great deal of the biological diversity in the plant and animal kingdoms is expressed in the morphology of organisms. In contrast, the range of morphological differences among microorganisms is relatively small. However, there is an impressive diversity of metabolism, and especially **catabolism**, in the microbial world. These alternative mechanisms for obtaining energy for growth are important for the proper function of the **biosphere**, and have practical implications for agriculture and industry.

CHAPTER NOTES

Energy can be captured from chemical compounds (**chemoorganotrophs**) or light energy (**photototrophs**). The discussion of metabolism earlier in the text considered the use of organic chemical compounds as energy sources. **Chemolithotrophic** bacteria can capture energy from the oxidation of inorganic chemicals. Under these conditions, carbon dioxide is usually the sole carbon source; most phototrophic microbes also grow **autotrophically** with CO_2 as the carbon source. The term autotroph means self feeding and indicates the organism can get all of its carbon needs from inorganic sources. In contrast heterotrophs require organic carbon and are dependent on the carbon fixing actions of the autotrophs.

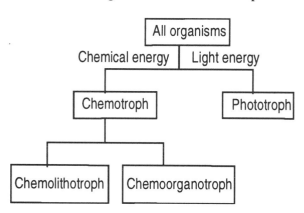

Light energy is converted into the chemical energy in the process of **photosynthesis**. The reactions are initiated by the absorption of a **quantum** of light by specific pigments; the **chlorophylls** are especially important. The chlorophyll converts light energy to chemical energy. The growth of phototrophs can be divided into the **light reactions** in which radiant energy is converted into chemical energy such as ATP, and **dark reactions** in which this energy is used to reduce CO_2 into cell material.

Chlorophyll molecules are **porphyrins**, and thus are related to the prosthetic group of cytochromes, but contain a magnesium atom rather than iron. The chlorophylls contain a long, hydrophobic side chain that causes the molecules to associate with membranes. There are a variety of chlorophylls that are chemically distinguished by the types of side chains on the porphyrin ring. These substituents also affect the wavelengths of light absorbed by the molecule. Thus, chlorophylls can be most easily distinguished by their **absorption spectrum**. **Chlorophyll a**, found in all plants, algae, and cyanobacteria, strongly absorbs blue and red light. Most cells have more than one pigment; in this way, a wider variety of wavelengths of light can be absorbed and used to drive photosynthesis.

The pigments are located in groups of a few hundred molecules in **photosynthetic membranes**. Some chlorophyll molecules are located in association with specific proteins and small molecules that are responsible for the subsequent conversion of light energy to chemical energy. These are **reaction center** chlorophylls. The surrounding **light harvesting** chlorophylls transfer any energy they absorb to the reaction center, and only then can chemical energy be generated.

In eukaryotic organisms, the photosynthetic membranes are located in specialized organelles, the **chloroplasts**. In prokaryotes, there is an extensive **internal membrane** system that contains the photosynthetic pigments.

ATP synthesis in photosynthesis bears several similarities to that in aerobic respiration. Both involve a series of **electron carriers** arranged in a membrane to generate a **proton motive force**. Electrons that travel through this electron transport system move from redox carriers of low redox potential to ones of higher potential, just as in aerobic respiration. The proton motive force is used to drive ATP synthesis by a membrane-bound ATPase. The difference between them is that in photosynthesis, the energy in light is used to drive a thermodynamically unfavorable reaction, the reduction of a molecule that does not have a strong tendency to accept electrons.

In **anoxygenic** photosynthesis, carried out by purple and green photosynthetic bacteria, ATP

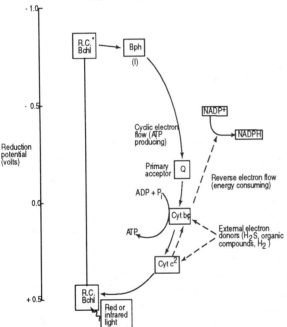

synthesis in photosynthesis is relatively straightforward. Oxygen is not generated during this type of photosynthesis. When the **special pair** bacteriochlorophyll in a reaction center is excited by light, they donate an electron to a **bacteriopheophytin**. This electron is then donated to a quinone and subsequently to a series of cytochromes, during which time protons are extruded across the membrane. Finally, the electron returns to the oxidized reaction center bacteriochlorophyll. Thus, the electron travels a cyclic path, and the process of ATP synthesis is called **cyclic photophosphorylation**.

Some of these anoxygenic phototrophic bacteria are autotrophs. To reduce CO_2 to organic compounds, both ATP and reduced NADP are necessary. However, the direct reduction of NADP is not thermodynamically possible by some of the **electron donors** that these microbes use. In these cases, the hydrolysis of ATP is used to force electrons from weak reductants to stronger reductants

that can then reduce NADP. This process, called **reversed electron transport**, is also used by lithotrophs to overcome the same problem.

When growing photosynthetically, anoxygenic phototrophic bacteria are strict anaerobes. However, some species can grow aerobically as **organotrophs**; O_2 inhibits the formation of bacteriochlorophyll in these species so they are incapable of photosynthesis under aerobic conditions. O_2 may interact with a regulatory protein and suppresses the formation of the photosynthetic gene cluster. These species often use organic compounds as their carbon source when photosynthesizing, and thus are **photoheterotrophs**.

In **oxygenic** photosynthesis, there are two distinct types of reaction center chlorophylls that operate in tandem. When the reaction center chlorophyll of photosystem II becomes oxidized by absorbing light, the electron that re-reduces it comes from water. There are two important points. (1) The electron that was removed from chlorophyll by light excitation can never return to it; therefore, the process of ATP synthesis that results from photosystem II activity is called **non-cyclic photophosphorylation**. (2) The end product of the oxidation of water is O_2.

As in cyclic photophosphorylation, the electron removed from reaction center II is passed through a series of electron carriers, and a proton motive force is generated which is used to drive ATP synthesis. However, the electron is finally donated to an oxidized reaction center chlorophyll of **photosystem I**. This molecule was oxidized by a photochemical reaction, which has generated a reductant that can reduce the protein **ferredoxin**. This protein can donate electrons to NADP+ to produce the reducing power necessary for CO_2 fixation. Alternatively, the primary electron acceptor of photosystem I can pass electrons through cytochromes back to the reaction center chlorophyll. ATP is synthesized as a consequence of these reactions; thus, the operation of photosystem I by itself results in cyclic photophosphorylation.

Pigments other than chlorophylls may serve as **accessory pigments**. These include **carotenoids**, which absorb blue light, and transfer some of the light energy to chlorophylls by **fluorescence**. **Phycobiliproteins** differ from other types of pigments in that they are water-soluble rather than associated with lipid, and they are macromolecules. A light-absorbing phycobilin is linked to a protein. Phycobiliproteins are found in cyanobacteria and red algae. The pigments are arranged in structures called **phycobilisomes** that are attached to the photosynthetic membrane. As a consequence of this arrangement, light energy absorbed by phycobiliproteins is transferred very efficiently to chlorophylls in the membrane.

Most autotrophs use a sequence of reactions termed the **Calvin cycle** to fix CO_2 and produce organic compounds. There are three phases to the cycle: (1) addition of CO_2 to an acceptor molecule, (2) reduction of this carbon to the oxidation level of cell material, and (3) regeneration of the acceptor molecule. Stage 1 is catalyzed by **ribulose bisphosphate carboxylase**; ribulose bisphosphate is the acceptor molecule. Stage 2 requires both ATP and reduced NADP. Stage 3 involves a number of sugar rearrangements to regenerate ribulose bisphosphate. To synthesize one molecule of glucose from six CO_2, twelve reduced NADP and eighteen ATP are expended.

Chemolithotrophs are prokaryotes that can gain energy by oxidizing **inorganic** compounds. There are several groups of chemolithotrophs that can be distinguished by the energy source that they use. However, there are several features in common. (1) When growing autotrophically, they use the Calvin cycle to fix CO_2. (2) Some are **facultative chemolithotrophs** that grow heterotrophically when appropriate organic compounds are available. (3) Some facultative chemolithotrophs will grow **mixotrophically**; that is, when inorganic and organic energy sources are present, they are used simultaneously. (4) Most chemolithotrophs are aerobes that synthesize ATP by oxidative phosphorylation. (5) With the exception of the H_2-oxidizing bacteria, lithotrophs use **reversed electron transport** to generate reducing power. The energy sources they use are not strong enough reductants to directly reduce NAD.

All **hydrogen bacteria** are facultative chemolithotrophs; a wide variety of bacterial genera have members that can use H_2 as an energy source. A membrane-bound or cytoplasmic **hydrogenase** transfers electrons from H_2 to an electron transport system.

The **colorless sulfur bacteria** oxidize reduced sulfur compounds to **sulfuric acid**, although **elemental sulfur** may be stored as an intermediate product. Some of these bacteria are acidophiles, and can reduce the pH of the medium to 2.

Iron-oxidizing bacteria such as *Thiobacillus ferrooxidans* is an **acidophilic** sulfur bacterium that can also use reduced iron as an energy source. This microbe uses the existing proton gradient between its pH-neutral cytoplasm and the extremely acid external environment to generate energy. When oxygen acts as a terminal electron acceptor to produce water, hydrogen ions are consumed. This reaction occurs on the cytoplasmic side of the membrane, so that protons are consumed in the cell. These protons are replenished from the proton-rich external environment by influx through the membrane-bound ATPase, thereby generating ATP.

Nitrogen oxidizing bacteria are able to use reduced forms of nitrogen, ammonia or nitrite, and oxidize it to either nitrite or nitrate, respectively. Theses processes are conducted by two species of **nitrifying** bacteria. The first step, oxidation of ammonia to nitrite is conducted by *Nitrosomonas* while the second step, oxidation of nitrite to nitrate is conducted by *Nitrobacter*.

Anaerobic respiration is a process in which an electron transport system is used to generate a proton motive force, but the terminal electron acceptor is a molecule other than oxygen. Most organisms that carry out anaerobic respiration are heterotrophs, and the terminal electron acceptors are generally inorganic compounds. However, there are exceptions to both these rules. The energy released (and therefore the amount of ATP synthesized) is less in anaerobic than in aerobic respiration because the difference in reduction potential between electron donor and acceptor is greatest when O_2 is the acceptor.

When oxidized inorganic compounds such as nitrate, sulfate, and carbon dioxide serve as terminal electron acceptors for catabolism, large quantities of reduced products are excreted into the medium. These compounds may also be reduced in the cell for use in biosynthesis. In this latter case of **assimilative metabolism**, relatively small amounts are reduced to meet the needs for growth, and the atoms are incorporated into cell material.

In **denitrification**, nitrate is reduced either to ammonia or to gaseous nitrogen compounds. The latter is of agricultural significance because it represents a decrease in the soil levels of nitrogen available for plant growth. All denitrifying bacteria are capable of aerobic respiration. However, when oxygen is absent, the membrane-bound proteins **nitrate reductase** and **nitrite reductase** can transfer electrons from the electron transport system to nitrate or nitrite.

The heterotrophic **sulfate-reducing bacteria** are obligate anaerobes in which H_2S is the final product of anaerobic respiration. Before sulfate can be reduced, it must be activated by reaction with ATP to form **adenosine phosphosulfate**. The enzyme **hydrogenase** is an integral part of energy conservation during the reduction of sulfate by the electron transport system. Hydrogenase on the outside face of the cytoplasmic membrane transfers electrons to cytochrome c_3 in the membrane; the electrons are used to reduce sulfate. However the hydrogen ions of H_2 remain outside the membrane; the result is a proton gradient that can drive ATP synthesis.

Methanogens and **homoacetogens** are obligate anaerobes in which CO_2 is the terminal electron acceptor. H_2 is the most common energy source used by these groups, and the products are methane and acetic acid, respectively.

In addition to nitrate, sulfate, and carbon dioxide, ferric iron (Fe^{3+}), manganese (Mn^{4+}) and several organic compounds can serve as electron acceptors for anaerobic respiration. The organic compounds that can serve as terminal electron acceptors include fumarate, glycine and trimethylamine oxide.

The oxidation of an organic compound to release energy generates two problems: (1) conserving some of the released energy and (2) disposing of the electrons. If an external electron acceptor is available (O_2, NO_3 etc.), both problems are solved by the use of an electron transport system which generates a proton motive force to drive ATP synthesis and which eventually donates the electrons to a terminal acceptor.

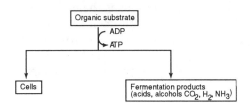

However, **fermentative** organisms generally must synthesize ATP by **substrate-level phosphorylation** in which a high-energy phosphate bond is transferred from one of ten high-energy intermediates to ADP. Furthermore, some of the intermediate organic products of metabolism must be used as electron acceptors. The quantity of oxidized

124

and reduced products in the fermentation must **balance**. Electron balance can be achieved by the production of H_2 by hydrogenase, using electrons donated by **ferredoxin**.

In anaerobic environments, the products of one fermentation are energy sources for other fermentative bacteria. The **terminal products** of anaerobic decomposition are CH_4 and CO_2. In some instances, the growth of one fermenter depends upon the utilization of its products by a second species. In these cases, the metabolism of the first microbe is thermodynamically unfavorable, unless the concentration of a fermentation product is kept low due to its consumption by another bacterium. The consumption of H_2 in **interspecies hydrogen transfer** is the best-studied example of this type of **syntrophic** association.

Most naturally occurring organic compounds can be metabolized aerobically or anaerobically by at least some microbial species. However, neither **lignin** nor **aliphatic hydrocarbons** are degraded under **anaerobic** conditions. Their stability has led to the formation of coal and oil deposits.

In the last fifty years, organic chemists have synthesized a number of organic compounds that did not exist in nature. Some of these **xenobiotic** compounds are not degraded by microbes, presumably because the opportunity for evolution of bacteria that can grow on these substances has not existed.

Sugars are the most common organic compounds. The bulk of them exist as insoluble **polysaccharides** such as cellulose. Therefore, microbes must excrete **exoenzymes** capable of hydrolyzing the macromolecule to smaller units.

Cytoplasmic polysaccharides that serve as energy reserves are hydrolyzed differently from extracellular ones. When a monomer is removed from a polysaccharide such as glycogen, it is **phosphorylated**. This saves energy, because the first step in glucose catabolism is generally the phosphorylation of glucose using ATP.

The microbial metabolism of sucrose plays a key role in initiating **tooth decay**. *Streptococcus mutans* uses the enzyme **dextran sucrase** to cleave the disaccharide sucrose. The fructose is used to generate energy, whereas the glucose monomer is used to synthesize **dextran**, a sticky extracellular polysaccharide that binds bacterial cells tightly to the tooth surface.

Organic acids that are intermediates of the TCA cycle can be used as carbon or energy sources by many microbes. This is because most microorganisms contain all or most of the enzymes of the **tricarboxylic acid (TCA) cycle**. However, when these intermediates are used for biosynthesis of cell material, the acceptor molecule **oxaloacetate** must be regenerated by other means. This can be accomplished by the enzymes isocitrate lyase and malate synthase, which together with some TCA cycle enzymes form the **glyoxylate cycle**. In effect, the two new reactions bypass the TCA cycle reactions in which the carbon from acetate is oxidized to CO_2. Alternatively, if pyruvate or phosphoenolpyruvate are in abundant supply, these can be **carboxylated** to form oxaloacetate.

Fats and lipids are hydrolyzed by **lipases** to glycerol and fatty acids. The fatty acids can be utilized by microbes via β-oxidation, in which two carbon units of **acetyl CoA** are sequentially removed from the long chain fatty acid. Acetyl CoA can then be metabolized in the TCA cycle.

Aliphatic and **aromatic** hydrocarbons often contain only carbon and hydrogen atoms, and no oxygen. The initial step of hydrocarbon metabolism often involves the introduction of oxygen atoms from O_2 by **oxygenase** enzymes. There are two types of these metal-containing enzymes. **Dioxygenases** incorporate both oxygen atoms of O_2 into the substrate. **Monooxygenases** incorporate one atom as a **hydroxyl** group; the other generally combines with protons and electrons from reduced NAD to form water. After this initial oxidation, hydrocarbons are converted into organic acids that can be metabolized in the TCA cycle.

Nitrogen for biosynthesis of cell material can be obtained from organic sources such as amino acids, or inorganic sources. Ammonia and nitrate are most commonly used by microbes, but some bacteria can reduce N_2 gas to synthesize organic nitrogen.

When ammonia concentrations are high, the enzyme **glutamate dehydrogenase** can be used to incorporate ammonia into organic compounds. However, at low concentrations, the **glutamine synthetase-glutamate synthase** system is more efficient. The net reaction converts alpha-ketoglutarate and ammonia to glutamate, with the consumption of one ATP. The **activity** of glutamine synthetase is controlled in response to ammonia concentration. At high concentration, the protein is **covalently modified** by **adenylylation**; this reduces its catalytic activity.

Nitrogen fixation is a process where some prokaryotic organisms reduce N_2 to ammonia, which is then incorporated into glutamate by the enzymes described above. A multimeric enzyme called **nitrogenase** catalyzes nitrogen fixation. One component of the complex is the protein **dinitrogenase**. Together with a cofactor containing iron and molybdenum atoms, this protein reduces nitrogen gas to ammonia. A second component, **dinitrogenase reductase** is necessary to transfer electrons from **ferredoxin** to nitrogenase. The reduction of one N_2 molecule requires the hydrolysis of 15-20 ATP molecules. ATP hydrolysis lowers the reduction potential of nitrogenase reductase sufficiently for the reaction to proceed. Eight electrons are consumed in the reaction, even though theoretically only six are required. The other two are lost in a molecule of H_2. Nitrogenase reductase is inactivated by O_2. Therefore, in aerobic bacteria, there must be oxygen-deficient microenvironments in the cell for nitrogen fixation to proceed. The activity of dinitrogenase is most conveniently assayed by adding acetylene. The enzyme reduces this triple-bonded molecule to form ethylene, which can be measured by gas chromatography.

SELF TESTS

COMPLETION

1. Some species of _____ can grow at pH values less than 3.

2. The glyoxylate cycle bypasses some of the reactions of the _____.

3. In β-oxidation, fatty acids are oxidized to _____, which enters the TCA cycle.

4. In _____, all ATP is synthesized by substrate-level phosphorylations.

5. The reaction center pigment in purple bacteria is _____.

6. The electron which re-reduces oxidized P700 of photosystem I is donated by _____.

7. _____ are the structures which contains photosynthetic membrane systems.

8. In oxygenic photosynthesis _____ photosystem(s) are necessary, and both _____ and _____ are produced for use in the Calvin cycle.

9. Bacteria that simultaneously use organic compounds and inorganic energy sources are _____.

10. Elemental sulfur serves as an energy reserve for _____ bacteria.

11. Chemolithotrophic bacteria synthesize ATP by _____.

12. Two unique electron carriers found in denitrifying bacteria are _____ and _____.

13. The sulfate-reducing bacteria activate sulfate by reacting it with _____.

14. Acetogenic bacteria produce acetate from _____ and _____.

15. Approximately _____ grams of cell dry weight can be produced per mole of ATP synthesized by a cell.

16. When two species can grow together under conditions in which neither alone could grow, their association is _____.

KEY WORDS AND PHRASES

β-galactosidase (p 521)	acetylene reduction (p 529)
amylase (p 520)	anoxygenic photosynthesis (p 481)
assimilative metabolism (p 503)	autotroph (p 477)
carotenoids (p 476)	cellulase (p 520)
chemolithotroph (p 476)	chloroplasts (p 480)
dissimilative metabolism (p 503)	electron acceptor (p 493)
electron donor (p 492)	endohydrolase (p 524)
fermentation (p 516)	fermentation balance (p 515)
heterolactic fermentation (p 519)	homolactic fermentation (P 516)
hydrolase (p 524)	invertase (p 521)
lactose (p 521)	mixed acid fermentation (p 516)
Na^+ translocating ATPase (p 518)	*nif* regulon (p 529)
Nitrobacter (p 499)	*Nitrosomonas* (p 499)
oxygenic photosynthesis (p 481)	P680 chlorophyll (p 486)
P700 chlorophyll (p 486)	phosphoribulokinase (p 489)
photosynthetic gene cluster (p 484)	plastocyanin (p 487)
rusticyanin (p 498)	sucrose (p 521)
sulfur bacteria (p 494)	superoperons (p 484)
thylakoids (p 480)	

MULTIPLE CHOICE

1. What type of microbial metabolism is used in the industrial production of alcohol, butanol, and acetone?

    (A) aerobic respiration

    (B) anaerobic respiration

    (C) chemolithotrophy

    (D) photosynthesis

    (E) fermentation

2. *Thiobacillus ferrooxidans* can use which of the following as an energy source?

    (A) sulfide

    (B) sulfate

    (C) ferrous iron

    (D) ferric iron

    (E) A and C above

3. Which of the following is NOT a characteristic of the enzyme nitrogenase?

    (A) requires ATP to carry out its reaction

    (B) requires reducing power to carry out its reaction

    (C) is inactivated by oxygen

    (D) is not found in aerobic bacteria

    (E) can reduce acetylene to ethylene

4. When cyanobacteria photosynthesize, they

    (A) synthesize ATP by cyclic photophosphorylation

    (B) use hydrogen sulfide to reduce chlorophyll

    (C) produce oxygen gas

    (D) are not inhibited by DCMU

    (E) do not contain an electron transport chain

5. In denitrification, nitrate is the

    (A) energy source

    (B) product excreted due to dissimilative metabolism

    (C) compound assimilated into cell material

    (D) electron donor

    (E) terminal electron acceptor

6. Lithotrophs and heterotrophs differ in their:

    1. sources of carbon

    2. capacity to utilize sunlight

    3. oxygen requirement

    4. sources of energy

    (A) 1,2 (B) 2,3 (C) 3,4 (D) 1,4 (E) 2,4

7. Oxidation-reduction reactions generally involve the loss or gain of:

    (A) ADP

    (B) ATP

    (C) electrons

    (D) sodium ions

    (E) water

8. The activity of the enzyme dinitrogenase will be limited by low concentrations of all the molecules listed below EXCEPT:

    (A) N_2

    (B) ethylene

    (C) H_2

    (D) ATP

    (E) reduced ferredoxin

9. The lithotrophic bacterium *Nitrosomonas* shares which of the following properties with the photosynthetic bacterium *Chromatium*?

    1. Grows anaerobically

    2. Uses the Calvin cycle

    3. Uses proton motive force coupled phosphorylation to synthesize ATP

    4. Often contain intracytoplasmic membranes

    (A) 1,2 (B) 2,3 (C) 1,4 (D) 2,3,4 (E) 1,2,3,4

10. Autotrophic CO_2 fixation via the Calvin cycle requires all of the following EXCEPT:

    (A) photosystem II

    (B) ribulose bisphosphate

    (C) ATP

    (D) reduced NAD or NADP

    (E) transaldolase and transketolase

11. Among organisms that carry out anaerobic respiration, which have the electron transport chain most similar to that of an obligate aerobe?

    (A) denitrifying bacteria

    (B) methanogenic bacteria

    (C) acetogenic bacteria

    (D) nitrifying bacteria

    (E) sulfate-reducing bacteria

12. The reduction of sulfate to sulfide by sulfate-reducing bacteria requires which of the following?

    (1) an electron transport chain

    (2) ATP

    (3) membrane-bound ATPase

    (4) O_2

    (5) an electron donor such as H_2

    (A) 1,5 (B) 2,3 (C) 2,4,5 (D) 1,3,5 (E) 1,2,5

13. Oxygenases are involved in aerobic catabolism of all the following carbon and energy sources **EXCEPT**:

    (A) aromatic compounds

    (B) saturated hydrocarbons

    (C) unsaturated hydrocarbons

    (D) volatile fatty acids

14. Fermentation differs from other forms of energy yielding metabolism in that:

    1. ATP synthesis in most fermenting organisms occurs primarily from substrate-level phosphorylation

    2. the substrate materials often serves as the final electron acceptor

    3. the reactions are inefficient and do not provide energy for growth

    4. in general the reactions result in the formation of only one product

    (A) 1 & 3; (B) 2 & 3; (C) 1 & 2; (D) 1, 2, & 4.

15. The fermentative conversion of ethanol to acetate by bacteria is considered to be thermodynamically unfavorable, requiring +19.36 kJ/reaction. However when this reaction takes place in the presence of second organism, that can use H_2, both reactions proceed. This is an example of:

    (A) interspecies hydrogen transfer

    (B) syntrophy

    (C) coupled fermentation

    (D) all of the above

    (E) only A and B

DISCUSSION

1. What kinds of pigments participate in photosynthesis? What characteristics make some of these pigments associate with lipids?

2. Distinguish between primary pigments and accessory pigments, and between antenna pigments and reaction center pigments.

3. Compare the mechanisms of ATP generation in phototrophic and heterotrophic microbes.

4. Compare the means by which reducing power is obtained in oxygenic and anoxygenic phototrophic microbes.

5. Compare and contrast cyclic and non-cyclic photophosphorylation with regard to (a) the number of different reaction centers involved, (b) the products of the reactions, and (c) the groups of microbes that use each mechanism.

6. How much ATP and reducing power are used in the Calvin cycle to reduce CO_2 to the oxidation level of cell biomass?

7. What enzymes are used in not only the Calvin cycle but also the (a) pentose phosphate cycle or (b) glycolysis?

8. Give two reasons why the growth yields of chemolithotrophic bacteria are low.

9. What molecules, other than O_2, can serve as terminal electron acceptors in oxidative phosphorylation?

10. Why are the enzymes isocitrate lyase and malate synthase specifically required when aerobic heterotrophs are growing on acetate?

11. Illustrate the metabolic steps involved in the beta-oxidation of volatile fatty acids.

12. Molecular oxygen is an initial reactant in the catabolism of what group of organic compounds?

13. How can the activity of nitrogenase be assayed?

14. List the components necessary to carry out the reduction of N_2 to ammonia.

15. How does denitrification differ from aerobic respiration with respect to (a) terminal electron acceptor, (b) electron donors, and (c) electron carriers in the respiratory chain?

16. Why must most chemolithotrophic bacteria expend ATP to generate reduced NAD from their energy source? What molecules are involved in the transfer of electrons from the energy source to NAD?

ANSWERS

Completion

1. *Thiobacillus*; 2. tricarboxylic acid cycle; 3. acetyl CoA; 4. fermentations; 5. bacteriochlorophyll a; 6. plastocyanin; 7. thylakoids; 8. 2, ATP, reduced NADP; 9. mixotrophs; 10. colorless sulfur bacteria; 11. oxidative phosphorylation; 12. nitrate reductase, nitrite reductase; 13. ATP; 14. H_2, CO_2; 15. ten; 16. syntrophic.

Multiple choice

1. E; 2. E; 3. D; 4. C; 5. E; 6. D; 7. C; 8. C; 9. D; 10. A; 11. A; 12. E; 13. D; 14. C; 15. E.

Discussion

1. See text Sections 13.2, 13.3 and 13.6

2. See text Sections 13.3 and 13.6

3. See text Sections 13.1, 13.4, 13.5 and 13.13

4. See text Sections 13.4 and 13.5

5. See text Sections 13.4 and 13.5

6. See text Section 13.7

7. See text Section 13.7

8. See text Sections 13.7, 13.8, and 13.11

9. See text Sections 13.14, 13.15, 13.13, 13.17, and 13.18

10. See text Section 13.22

11. See text Section 13.22

12. See text Section 13.24

13. See text Section 13.25

14. See text Section 13.25

15. See text Section 13.15

13. See text Sections 13.4, 13.8, and 13.11

Chapter 14
MICROBIAL ECOLOGY

OVERVIEW

Chapter 14 (pages 534-607) centers on the diversity in survival and growth strategies that allow microbes to exist in many different environments. Microorganisms are responsible for many reactions essential for the proper functioning of the biosphere. In natural environments, the interaction of a microbe with the physical and chemical characteristics of the habitat, as well as with other organisms, will determine its success in growing there. The **recycling** of nutrients from organic compounds into inorganic forms that can be used by photosynthetic organisms is an especially important function performed by microbial **decomposers**.

CHAPTER NOTES

Microbial ecologists are fundamentally interested in knowing which microbes are present and what they are doing. Microbial ecology is a broad and varied field, but it is possible to make some generalizations. (1) The physical and chemical conditions in the **microenvironment** to which microscopic organisms are actually exposed may be different from what we can conveniently measure in the size of sample we collect. (2) Nutrient availability limits growth rate in most natural environments. Furthermore, the supply of nutrients may not be continuous, but occurs in pulses, so that growth occurs in spurts with intervening periods of starvation conditions. (3) Microbes are often found attached to **surfaces** because in nutrient-limited environments the nutrient concentrations are higher there. (4) Microbial species compete for the limiting nutrient in the environment. The successful species is the one that can grow fastest at low nutrient concentrations, unless a competitor can produce a substance that inhibits growth.

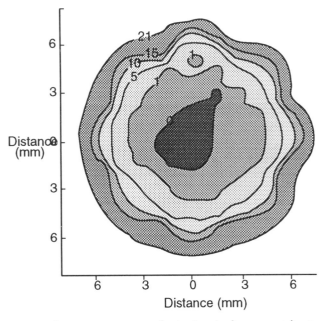

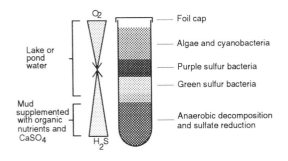

Enrichment cultures are used to select microbes capable of particular metabolic transformations. This involves using the proper **selective** conditions in the medium, and using an **inoculum** that contains the microbe of interest. The **Winogradsky column** is an example of an enrichment culture in which the initial set of microbial activities creates conditions that select for a variety of other microbes. Chemostats can be used to enrich for microbes that function best at low substrate concentrations.

The number of microbes in a natural sample can be determined by cultural methods, such as the **plate count**, or by direct microscopic count. In the latter case, the microbes are often visualized by using fluorescent dyes such as **acridine orange** that stain all cells, or by using **fluorescent antibodies** or **nucleic acid probes** to specifically identify a single species. Enumeration methods that involve culturing organisms suffer from the problem that not all viable bacteria will grow on a specific medium. A second problem associated with all enumeration methods is that they do not indicate whether the microbes are actually active in the environment. A more recent approach to understanding the makeup and structure of microbial communities involves the direct extraction of DNA (or RNA) from a natural sample. This recovered genetic material is characterized and the community is then described. This approach avoids the changes that occur in a population as a result of sub-culturing onto nutrient rich medium.

A variety of techniques are used to measure microbial activity. The problem is often that microbial population sizes are small, so sensitive methods are needed. ATP measurements are sensitive indicators of living biomass in a sample. **Radioisotopes** provide a sensitive means of measuring specific metabolic reactions mediated by microbes. The chemical conditions in microenvironments can be monitored using **microelectrodes**; these have been especially useful in studying **microbial mats**.

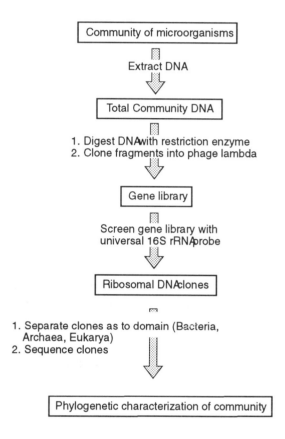

The amount of a **stable isotope** in geological deposits can be used to determine whether the deposits were a result of biological activity. Most enzymatic reactions will slightly favor the use of a lighter isotope if it is available. Analysis of the ratio between two stable isotopes of a chemical element in ancient sediments is also useful to determine how long ago certain biological processes evolved.

The production of organic matter in aquatic habitats is primarily due to photosynthetic microorganisms, the **primary producers**. The rate of primary production is determined by the physical environment, and the supply of inorganic nutrients. For example, there are few nutrients available in the open ocean and primary production is generally very low there in comparison to near-shore areas.

Aquatic environments that develop vertical density gradients due to **thermal stratification** may become anaerobic in the bottom stratum. Organic matter

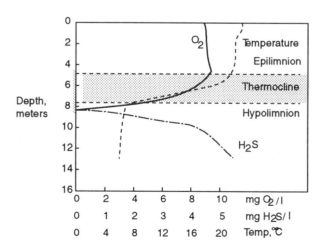

134

produced through photosynthesis in the surface waters sinks to the bottom, where heterotrophic bacteria consume O_2 as they mineralize it. If enough organic matter sediments to the bottom, all oxygen is consumed by this process, and further chemoheterotrophic metabolism must use anaerobic processes. Anaerobic conditions exclude most higher animals from the stratum.

The formation of soil involves physical, chemical, and biological processes. Microbes contribute by producing organic matter, which is converted to carbonic acid or organic acids that dissolve rock minerals. The formation of soil from exposed rock may take hundreds of years. Microbes in soil control the availability of many important plant nutrients. As a result the microbial functioning in soil is a key part of that soils productivity. Microbial activity in soil is limited by **water activity** (under both wet and dry conditions) and nutrient status. Recent work has suggested that microbes can exist several hundred meters beneath the earth's surface. It is unclear as to the level of activity in these subsurface environments.

Primary production in the open ocean is low. Therefore, only small amounts of organic matter reach the ocean floor at depths greater than 1000 m. In addition, the temperature in these regions is less than 4°C, and hydrostatic pressure exceeds 100 atmospheres. As a consequence, there is little biological activity in these regions.

There are some regions on the ocean floor of high biological activity. These are **hydrothermal vents** from which hot, mineral-rich water emanates. The water contains reduced sulfur compounds that serve as energy sources for **S-oxidizing chemolithotrophs**. These autotrophic bacteria perform the function fulfilled by photosynthetic organisms on the earth's surface -- they are the primary producers of the ecosystem. Dense populations of invertebrate animals are found near the vents. However, they do not eat the sulfur bacteria, but rather form symbiotic associations with them. The tube worm *Riftia* contains a modified gastrointestinal tract called the **trophosome** that is filled with these microbes. The sulfur-oxidizing bacteria are located in the gill tissue of clams and mussels. The animals can then live on excreted organic matter and dead bacterial cells. The tube worms contain an unusual **hemoglobin** that can transport H_2S to the bacteria in the trophosome without it poisoning the animal. From studies of the source water from black smokers, which can be as hot as 380°C, the upper temperature limit for life seems to be about 150°C.

Transformations of **carbon** drive the cycles of all other elements because they require energy provided either by photosynthesis or the catabolism of organic compounds. The most important fluxes of carbon are through atmospheric CO_2 and living and dead organic matter. CO_2 fixation by photosynthesis and its production by respiration are the mechanisms by which these fluxes occur.

Anaerobic decomposition of organic carbon requires the participation of four metabolic groups of bacteria. (1) Bacteria with extracellular hydrolyses cleave high molecular weight polymers such as polysaccharides or proteins into their constituent monomers. (2) Fermentative bacteria convert these monomers to organic acids, H_2, and CO_2. (3) **Fatty acid oxidizing bacteria** convert the organic acids to acetate, H_2, and CO_2. (4) These compounds are used as substrates by **methanogenic bacteria** to produce CH_4. Therefore, the end products of anaerobic decomposition are the gases CH_4 and CO_2.

The fatty acid oxidizing bacteria can only grow if the H_2 that they produce is consumed in another reaction. These organisms depend upon **interspecies hydrogen transfer** in which other bacteria such as the methanogens quickly metabolize any H_2 that they generate.

Methanogenesis occurs in anaerobic freshwater environments such as lake sediments, marshes, and swamps, as well as in the rumen. In anaerobic environments with high sulfate concentrations (such as marine habitats), sulfate-reducing bacteria out compete methanogens for H_2 and acetate, and these bacteria are responsible for the terminal reactions of anaerobic decomposition.

Ruminant animals contain in their gastrointestinal tract an anaerobic fermentation chamber, the **rumen**. Microbes digest cellulose or other polysaccharides to **volatile fatty acids** that are assimilated through the wall of the rumen and used by the animal as an energy source. The gases carbon dioxide and methane are the other major products of microbial metabolism. The microbes that grow in the rumen can subsequently be digested by the animal and used as a source of amino acids and vitamins.

The microbes in the rumen include bacteria and **protozoa** that hydrolyze cellulose to glucose. A variety of fermentative anaerobes metabolize the glucose or metabolites of glucose to organic acids, H_2, and CO_2. Methanogenic bacteria convert the latter two products to methane gas. As mentioned above, organic acids (acetate, propionate, and butyrate) are assimilated by the animal.

The element **nitrogen** exists in a number of oxidation states. Several redox transformations involving nitrogen compounds are mediated exclusively by microbes. N_2 in the atmosphere is the major reservoir of nitrogen. Some prokaryotes are capable of **nitrogen fixation**, in which N_2 is reduced to ammonia in an energy-demanding process. Nitrogen recycling generally involves ammonia and nitrate. Although N_2 reduction can occur chemically, 85% of nitrogen fixation is due to microbial activity.

Denitrification is the primary route for biological production of gaseous nitrogen. This process represents a decrease in nitrogen available for plant growth.

The decomposition of organic matter results in **ammonification**, the release of ammonia. Under anaerobic conditions ammonia accumulates, but in aerobic environments plants will quickly incorporate it into amino acids.

Chemolithotrophic bacteria can transform ammonia to nitrate in aerobic environments, in a process called **nitrification**. The nitrate is leached more easily from soil than ammonia; therefore, the process is not agriculturally beneficial.

Sulfur, like nitrogen, can occur in a variety of oxidation states. However, some redox transformations occur chemically, in the absence of biological catalysis. The three oxidation states of significance are sulfide (sulfhydryl), elemental sulfur, and sulfate.

Sulfate is used as a sulfur source by many organisms; in doing so they reduce sulfate to sulfhydryl groups in organic sulfur compounds. **Sulfate-reducing bacteria** carry out the **dissimilative** reduction of sulfate to hydrogen sulfide in anaerobic environments. These organisms use organic compounds or H_2 as a source of electrons to reduce sulfate.

Sulfide is chemically oxidized by O_2. Biological sulfide oxidation by chemolithotrophic bacteria occurs in strata where vertical gradients of sulfide and O_2 overlap. Under anaerobic conditions, photosynthetic sulfur bacteria oxidize sulfide.

The form of iron in natural environments is influenced by pH and oxygen. Oxygen will chemically oxidize ferrous iron to ferric iron. This form is only soluble at acidic pH. However, ferric iron can be kept in solution by forming a complex with organic compounds. Ferric iron can be reduced chemically by hydrogen sulfide, or microbiologically.

At acid pH, *Thiobacillus ferrooxidans* uses ferrous iron as its energy source, and produces ferric iron. This reaction is of great importance in the formation of acid leachate from mining operations, and in the microbial leaching of metals from ores. Both processes involve microbiological and chemical reactions. Under **aerobic** and acidic conditions, *T. ferrooxidans* oxidizes ferrous to ferric iron. Ferric iron chemically oxidizes **pyrite** to form more ferrous iron and sulfuric acid.

Acid mine drainage can have a pH as low as 2. The acidic conditions solubilize metals such as **aluminum**; both the pH and metal concentrations are toxic to aquatic life. Acid mine drainage can be prevented by covering pyrite minerals to prevent exposure to air; *T. ferrooxidans* requires O_2 to oxidize ferrous iron.

In microbial leaching, the chemical and microbiological oxidation of pyrite generate ferric ions that chemically oxidize metal ores and thereby solubilize the metal. The process is used commercially to recover **copper** and **uranium** from low-grade ores.

Mercuric ion (Hg^{+2}) is toxic to organisms because it reacts with sulfhydryl groups in proteins. Some bacteria detoxify mercuric ion by converting it to **methylmercury** or **dimethylmercury**. In the environment, both these compounds become concentrated in the **lipids** of animals, where they are more toxic than mercuric ion itself.

The hydrocarbons in petroleum are attacked by bacteria and yeasts under aerobic conditions. Under anaerobic conditions, petroleum deposits are not decomposed. Natural gas (methane) is used as an energy source by **methanotrophic** bacteria under aerobic conditions, but appears to be stable in anaerobic environments.

Synthetic chemicals, the **xenobiotics**, vary in their susceptibility to microbial attack. Small changes in chemical structure can have large effects upon **biodegradation**, as in the case of 2,4-D and 2,4,5-T. The eventual loss of these compounds from the environment is desirable, so that toxic chemicals do not accumulate. Xenobiotic loss can result from volatilization, spontaneous chemical reactions, leaching or biodegradation.

Many synthetic chemicals have only been present in the environment for 50 years, yet it is possible to find microbes that degrade them. The rapid evolution of such microbes may be due in part to the fact that the genes specifying the pertinent enzymes are often located on mobile genetic elements such as plasmids and transposons. Thus, new combinations of these genes can be constructed in bacteria via genetic exchange mechanisms. These locations also make it feasible to genetically engineer bacteria with the ability to degrade specific toxic molecules.

Most microbial activity associated with plants occurs in the **rhizosphere**, the region surrounding the roots. Organic compounds that stimulate heterotrophic microbes are excreted through the roots.

Plants have several defense mechanisms to prevent microbial infection. (1) Toxic chemicals such as cyanide and sulfide are produced nonspecifically. (2) Aromatic compounds called **phytoalexins** are produced in response to a microbial infection. (3) A **hypersensitive response** kills plant cells in the vicinity of an infected cell to prevent the spread of infection.

Agrobacterium species induce the uncontrolled division of plant cells to form tumor-like growths. To induce the formation of malignant **callus**, the bacterium must infect a wound on the plant. After induction, the presence of *Agrobacterium* is unnecessary. This is because the induction process entails the transfer of **T-DNA** from the bacterial **Ti plasmid** to a plant cell where the T-DNA is integrated into the plant genome. T-DNA contains genes for tumor formation, and for biosynthesis of unusual amino acids, the **opines**.

The Ti plasmid in the bacterium contains genes for (1) using the opines produced by the plant tumor as energy sources, (2) producing two plant growth hormones, **auxin** and **cytokinin**, which stimulate tumor development, and (3) increasing the virulence of *Agrobacterium*.

The Ti plasmid can be used as a **vector** to introduce foreign DNA into plant cells, using the techniques of genetic engineering. In this way, desirable traits, such as herbicide resistance and resistance to drought, salinity, or pathogens could be added to specific plant species. To be useful as a vector, the tumor functions of the Ti plasmid must be inactivated, and suitable means to regenerate whole plants from single cells must be developed.

Species of *Rhizobium* and *Bradyrhizobium* make specific associations with **leguminous** plants to form **root nodules** capable of **nitrogen fixation**. Because the availability of combined nitrogen in soil often limits the growth of agricultural crops, these associations are economically important.

In the root nodule, it is the bacterium that contains the enzyme **nitrogenase**. This enzyme is not normally produced by the cells when they are not in root nodules. The O_2 level in the nodule is controlled by a plant protein, **leghemoglobin**. This control is necessary because the bacteroids carry out aerobic respiration to provide energy for nitrogen fixation, but nitrogenase is inactivated by oxygen.

Plant species are nodulated by particular species of *Rhizobium* or *Bradyrhizobium*. This specificity resides in the early stages of nodule formation, when the bacterium **attaches** to the root hair of the plant. Recognition is mediated by plant **lectins**, which bind to a bacterial surface polysaccharide.

After specific binding to the root hair tip, nodulation involves (1) entry of the bacterium into the root hair and formation of an **infection thread**, (2) travel to the main root and infection of **tetraploid** plant cells, and (3) repeated division of the tetraploid cells and transformation of the bacterial cells into **bacteroids** capable of nitrogen fixation.

The plant supplies organic acids produced via photosynthesis to the bacteroid across the **peribacteroid membrane**. These are used to synthesize ATP and reducing power necessary for the reduction of N_2 to ammonia. In return, the ammonia generated by nitrogenase is incorporated by the plant into **glutamine** and other organic nitrogen compounds. These are translocated to plant tissues for use in growth.

There are other associations between plants and nitrogen-fixing microbes. In rice paddies, the fern *Azolla* contains cyanobacteria that possess specialized cells called **heterocysts** that fix nitrogen. Alder trees contain root nodules with the actinomycete *Frankia*. A looser association occurs between some tropical grasses and the bacterium *Azospirillum lipoferum*.

SELF TEST

COMPLETION

1. In ecosystems, _____ are recycled, but there must be continual inputs of _____.

2. The set of physical and chemical conditions experienced by a microorganism is the _____.

3. The bacteria attached to soil particles can be visualized for microscopy by using _____.

4. Sensitive and specific measurements of the rate of chemical transformations can be made using _____.

5. Microbial mats are often found in _____ and _____.

6. Free-floating primary producers in aquatic systems are called _____.

7. The oxygen in the bottom layers of a lake is consumed by _____ microorganisms.

8. _____ is a crude measure of the amount of organic matter in a body of water.

9. Highly productive regions on the deep ocean floor are associated with _____.

10. The gastrointestinal tract of the tube worm which lives near hydrothermal vents contains _____.

11. Carbon dioxide is removed from the atmosphere as a consequence of _____, and is returned due to _____.

12. The symbiotic relationship between methanogens and fatty acid-oxidizing bacteria is based upon _____.

13. _____ serve as the protein and vitamin source in the diet of ruminant animals.

14. The conversion of ammonia to nitrate by chemolithotrophic bacteria only occurs when oxygen is _____.

15. The reservoir of most of the sulfur on earth is _____.

16. The reduction of sulfate to hydrogen sulfide occurs in the _____ of oxygen gas.

17. Under aerobic conditions at pH 7, most iron atoms would be in the form of _____.

18. The oxidation of pyrite associated with coal ore leads to the formation of _____.

19. Xenobiotic compounds that do not serve as carbon and energy sources for bacteria may still be microbiologically transformed via _____.

20. The highest levels of microbial activity associated with plants occur in the _____.

21. Alder trees possess nitrogen-fixing root nodules which contain the bacterium _____.

KEY WORDS AND PHRASES

acid mine drainage (p 535)	benthic algae (p 552)
biodegradable polymer (p 590)	biogeochemical cycle (p 573)
co-metabolism (p 535)	decomposers (p 565)
ecosystem (p 535)	effective root nodules (p 600
enrichment culture (p 539)	herbicide (p 586)
ineffective root nodules (p 600)	jarosite (p 578)
lichens (p 591)	microbial ore leaching (p 580)
microbial plastics (p 590)	mineral soils (p 554)
mycorrhizae (p 592)	nod genes (p 601)
organic soils (p 554)	persistence (p 586)
phyllosphere (591)	phytoplankton (p 552)
radioisotope (p 549)	reductive dechlorination (p 587)
rhizosphere (p 551)	soil microenvironment (p 537)
soil organic matter (p 554)	soil profile (p 537)
stable isotope (p 549)	stem-nodulating rhizobia (p 597)

sub-surface soils (p 555)	Sym plasmid (p 600)
Winogradsky column (p 540)	yellow boy (p 578)

MULTIPLE CHOICE

1. What is the primary role of organotrophic bacteria in ecosystems?

    (A) nitrogen fixation

    (B) heavy metal detoxification

    (C) decomposers

    (D) primary producers

    (E) organic soil formation

2. Associations between two organisms which benefit each of them are:

    (A) antagonistic

    (B) antibiotic

    (C) parasitic

    (D) symbiotic

    (E) none of the above

3. All of the following groups of microorganisms would be found below the surface of a Winogradsky column EXCEPT:

    (A) fermentative heterotrophs

    (B) purple sulfur bacteria

    (C) sulfate reducing bacteria

    (D) aerobic heterotrophs

    (E) green sulfur bacteria

4. Viable counting methods provide an inaccurate picture of the number of microbes in a natural sample because:

    (A) the number of active bacteria is overestimated

    (B) no single medium is adequate to grow all bacteria

    (C) the culturing technique may induce dormancy

    (D) the methods are not very sensitive

    (E) many bacteria die as soon as they are removed from their habitat.

5. Which of the following can be used to estimate phytoplankton biomass in natural samples?

    (A) microelectrodes

    (B) chlorophyll

    (C) radioactive carbon dioxide

    (D) stable isotopes of carbon

    (E) fluorescent antibodies

6. If two isotopic forms of an element are available, enzymes are more likely to react with molecules that contain the

    (A) lighter isotope

    (B) heavier isotope

7. Rank the following aquatic habitats in order of INCREASING rates of primary production.

    1. lakes

    2. open ocean

    3. areas of upwelling in ocean

    (A) 1,2,3 (B) 2,3,1 (C) 3,2,1 (D) 1,3,2 (E) 2,1,3

8. The major factor(s) affecting microbial activity in soil is the availability of

    (A) water   (D) A and B above

    (B) nutrients   (E) B and C above

    (C) oxygen

9. The energy source which drives the hydrothermal vent ecosystem is

    (A) light

    (B) heat

    (C) reduced sulfur compounds

    (D) organic compounds

    (E) hydrogen gas

10. Which of the following compounds are important substrates for methanogenic bacteria in natural environments?

    (A) methane

    (B) hydrogen gas

    (C) acetate

    (D) propionate

    (E) B and C above

11. Much of the H_2 used by methanogens in anaerobic environments is produced by bacteria which metabolize

    (A) carbohydrates

    (B) fatty acids

    (C) proteins

    (D) cellulose

12. All of the following are important groups of microbes in the rumen EXCEPT:

    (A) cellulose degraders

    (B) methanogens

    (C) sulfate-reducing bacteria

    (D) carbohydrate fermenters

    (E) cellulolytic protozoa

13. The formation of nitrogen gas from nitrate by microorganisms is called

    (A) nitrogen fixation

    (B) nitrification

    (C) denitrification

    (D) ammonification

    (E) nitrogen assimilation

14. The oxidation of pyrite requires all of the following EXCEPT:

    (A) *Thiobacillus ferrooxidans*

    (B) oxygen gas

    (C) chemical oxidation of pyrite by ferric iron

    (D) acidic conditions

    (E) elemental sulfur

15. All of the following are NOT susceptible to microbial degradation EXCEPT:

    (A) methane, under anaerobic conditions

    (B) petroleum, under anaerobic conditions

    (C) cellulose, under anaerobic conditions

    (D) chlordane

16. In both *Agrobacterium* and *Rhizobium*, genetic information required for association with plants is contained on

    (A) lysogenic bacteriophage

    (B) insertion elements

    (C) plasmids

    (D) transposons

    (E) spores

DISCUSSION

1. What physical, chemical, and biological processes are involved in the formation of soil from rock?

2. Why do the bottom layers of some bodies of water become depleted of oxygen?

3. What methods are used to measure concentration gradients of oxygen, pH, and hydrogen sulfide that occur over distances of mm?

4. Explain why sulfate-reducing bacteria are responsible for the terminal reactions of anaerobic decomposition in marine environments rather than methanogenic bacteria.

5. Discuss the reactions involved in the formation of acid mine drainage. How can acid runoff from mining sites be prevented?

6. What mechanisms do microorganisms use to detoxify mercury compounds?

7. Why do some herbicides and pesticides persist for extended periods of time in soil?

8. Discuss three mechanisms by which plants defend themselves against microbial infections.

9. What steps are involved in the formation of a crown gall tumor in a plant. Which of these steps require the presence of *Agrobacterium tumefaciens*?

10. What steps are involved in the formation of a root nodule in a legume. Which of these steps require the presence of bacteria?

ANSWERS

Completion

1. nutrients, energy; 2. microenvironment; 3. fluorescent dyes; 4. radioisotopes; 5. hot springs and shallow marine basins; 6. phytoplankton; 7. organotrophic; 8. Biochemical oxygen demand; 9. hydrothermal vents; 10. symbiotic bacteria; 11. photosynthesis, respiration; 12. interspecies hydrogen transfer; 13. rumen microbes; 14. present; 15. sediments and rocks; 16. absence; 17. insoluble ferric hydroxide; 18. acid mine drainage; 19. co-metabolism; 20. rhizosphere; 21. *Frankia*.

Multiple choice

1. C; 2. D; 3. D; 4. B; 5. B; 6. A; 7. B; 8. D; 9. C; 10. E; 11. B; 12. C; 13. C; 14. E; 15. C; 16. C.

Discussion

1. See text Section 14 .8

2. See text Section 14 .7

3. See text Section 14 .5

4. See text Section 14 .12

5. See text Section 14 .14

6. See text Section 14 .18

7. See text Section 14 .20

8. See text Section 14 .22

9. See text Section 14 .23

10. See text Section 14 .23

Chapter 15
MICROBIAL EVOLUTION, SYSTEMATICS, AND TAXONOMY

OVERVIEW

Chapter 15 (pages 608-637) focuses on the genetic diversity found within the groups of microorganisms and how the diversity is thought to have arisen. It is clear that few, if any, locations on the earth's surface are not colonized by bacteria. Changes in the genetic constitution of microbes can lead to the **evolution** of new types of microorganisms. Although most genetic alterations are deleterious to the microbe, some changes that confer new properties on the cell may be beneficial in specific **environments**. Therefore, the interaction between genetic variability and environmental selection has produced the diversity of microbes that have developed during the earth's history.

CHAPTER NOTES

The earth is about 4.6 billion years old. At that time, the earth must have been very hot, but there is evidence that liquid water was present 3.8 billion years ago. Microbial **fossils** have been found in rocks less than or equal to 3.5 billion years old.

At this time, there was no O_2 in the atmosphere, so the first microbes must have been anaerobes. Furthermore, the earth was hotter than at present, so they were probably thermophilic. It is thought that a variety of organic compounds, including polymers, were formed by chemical reactions driven by ultraviolet radiation and lightning discharges.

The aggregation of polymeric molecules is hypothesized to have given rise to a cell capable of **metabolism** and **self-replication**. The rich supply of organic compounds present in the environment were used for energy generation, probably by **fermentation**. These could also be used for biosynthesis. Subsequent mutation and selection would yield new organisms with greater biosynthetic capacities.

The evolution of **porphyrins** was probably a key step, because, with these, electron transport phosphorylation could occur, so that for the first time non fermentable organic compounds could be used as energy sources, via **anaerobic respiration**.

The next step may have been development of photosynthetic pigments, such as the **chlorophylls**. Using anoxygenic photosynthesis, organisms could arise which were not dependent for energy upon the organic compounds produced by chemical reactions.

The occurrence of **oxygenic photosynthesis** fundamentally changed the earth and its evolution. The accumulation of atmospheric O_2 led to the formation of the **ozone** barrier to prevent intense ultraviolet radiation from reaching the earth. This meant organisms could exist over the entire surface of the earth. The presence of oxygen provided the conditions for the evolution of aerobic prokaryotes, eukaryotic cells, and subsequently the metazoans, higher animals, and plants.

For over 80% of the time during which life has existed on earth, that life consisted solely of microorganisms.

From the first cell that originated life on earth, three lines of descent were established. These are the **Bacteria**, **Archaea**, and **Eukarya** (See Figures 15. 7 and 15.12). Present-day eukaryotic cells are descended from the Eukarya line. The cytoplasmic organelles, mitochondria

144

and chloroplasts, found in eukaryotic cells were derived from prokaryotic **endosymbionts** (an aerobic heterotroph and a phototroph, respectively) which entered cells of the nuclear line while the Eukarya were evolving. The endosymbiont supplied the eukaryote with energy, and in return received nutrients and a protected environment. With time, the endosymbiont lost the genetic capability to exist independently.

Phylogeny is the study of the evolutionary relationships among organisms. The phenotypic characteristics of microbes have provided little information on microbial phylogeny. Recently, sequence comparisons between macromolecules of homologous function in different species have permitted an analysis of evolutionary distance. The study of **16S ribosomal RNA** has been exceptionally useful in this regard.

This method presumes that the longer the period of time since two organisms had a common ancestor, the greater the number of differences in sequence between a macromolecule of similar function which they both contain. The macromolecule chosen for study should be broadly distributed among organisms, have the same function in each, and not evolve so rapidly that similarities between distantly related organisms cannot be recognized.

Ribosomal RNA molecules are extremely well suited for this type of analysis. In addition to their universal distribution and constant function, they are easily purified from cells and can now be easily sequenced, using **DNA primers** to highly conserved sequences and the enzyme **reverse transcriptase**. Furthermore, some regions of the molecule have evolved rapidly while others have changed more slowly; this permits evolutionary distances to be calculated between both closely and distantly related species.

Phylogenetic trees are constructed from 16S RNA sequences by computer analysis of the sequence differences between each pair of organisms for which data exist. The computer algorithm arranges the organisms on a branching tree, spaced in a way to provide the best fit to all of the pair-wise sequence comparisons.

The 16S RNA sequences have allowed analysts to identify **signature sequences** that are unique to particular groups of organisms. In the future, as more are identified, these sequences may become useful for identifying unknown organisms.

The analysis of microbial phylogeny derived from ribosomal RNA sequences has led to many interesting insights. These include the following. (1) There are two lines of prokaryotic cells, the Bacteria and Archaea, which are no more closely related to each other than they are to the Eukarya line. (2) Different lineages have evolved at different rates; for example, the Archaea have evolved relatively slowly and the eukaryotic line has evolved rapidly. (3) Eukaryotic mitochondria and chloroplasts arose from endosymbiotic bacteria.

Bacteria can be divided into twelve groups that have been defined on the basis of ribosomal RNA analysis (rRNA). Most groups (similar to phylums) contain a variety of physiological and morphological types of bacteria. This reinforces the idea that phenotypic characteristics are inadequate to define evolutionary relationships between microbial species.

There are three groups of **Archaea**, two of which contain **methanogens**, while another contains **sulfur-dependent** organisms. This last group is composed of extreme thermophiles that require elemental sulfur for optimal growth. For most members, the sulfur serves as an electron acceptor in anaerobic respiration. Evolution of the eukaryotic line was characterized by periods of rapid evolution interspersed with eras of slow evolution. The accumulation of O_2 in the atmosphere about 1.5 billion years ago seems to correspond to a period of rapid evolution.

Differences among the primary kingdoms: The cell walls of Bacteria contain **peptidoglycan**. Most Archaea have glycoprotein in their cell walls. Eukaryotic cell walls (when present) have neither of these compounds but may contain cellulose or chitin.

The lipids of Archaea are unique in that they are **ether-linked** molecules, whereas in Bacteria and eukaryotes there are **ester links** between glycerol and the fatty acids.

The types and subunit complexity of **RNA polymerase** differ among organisms in the three kingdoms. Bacteria have a single RNA polymerase that consists of four different subunits. Archaea have at least two types of RNA polymerase, and these enzymes contain 8-10 polypeptides. Eukaryotic cells contain at least three different polymerases; the most prevalent type contains 10-12 polypeptides.

Protein synthesis in Bacteria and Archaea occurs on 70S ribosomes, whereas eukaryotic ribosomes are larger. **Formylmethionine** is always the first amino acid incorporated into Bacterial proteins; in eukaryotes and Archaea, an unmodified methionine is inserted by the initiator tRNA. Inhibitors of Bacterial protein synthesis generally do not affect the process in organisms from the other two kingdoms. Conversely, diphtheria toxin inhibits protein synthesis in all organisms except Bacteria.

Bacterial **taxonomy** relies on phenotypic characteristics to classify organisms, and is useful for the practical identification of unknown strains. The primary taxonomic unit is the **species**, which is defined by the phenotypic characteristics of a collection of similar strains. Culture collections contain **type strains** to serve as standards of the characteristics attributed to a particular species.

In conventional taxonomy, some characteristics are given special emphasis. These include the Gram stain, cell morphology, and the presence of cell structures such as endospores. In **numerical taxonomy**, all phenotypic characteristics are given equal weight in classifying strains.

Bergey's Manual of Systematic Bacteriology contains the phenotypic characteristics used to classify bacteria by conventional taxonomy, and **keys** that can be used to identify unknown strains from their phenotypic characters.

Some analyses of nucleic acids have been used in conventional taxonomy. These include measurements of **DNA base composition** and **nucleic acid hybridization**.

DNA base composition can only prove that organisms are unrelated. The ratio of bases in DNA can vary over a wide range. If two organisms have different DNA base compositions, they are not related. However, organisms with identical base ratios are not necessarily related, because the nucleotide sequences in the two organisms could be completely different.

Hybridization between the total DNA of two organisms is useful for detecting relationships between closely related organisms. Different genera rarely exhibit any DNA sequence homology. Strains of the same species should have homology values above 60-70 percent.

SELF TESTS

COMPLETION

1. The age of the earth has been estimated from _____ measurements.

2. The minimum properties which the first organism must have possessed were _____ and _____ .

3. When mechanisms to synthesize porphyrins evolved, it would have been possible for organisms to contain _____ .

4. The _____ theory maintains that eukaryotic organelles were derived from prokaryotic cells.

5. _____ and _____ are classes of molecules which can be used as evolutionary chronometers.

6. Oligonucleotide stretches of ribosomal RNA which are unique to specific groups of microorganisms are called _____.

7. Peptidoglycan is only found in _____.

8. The lipids found in Bacteria have more in common with those of _____ than those found in _____.

9. The basic taxonomic unit in microbiology is _____.

10. To achieve formal taxonomic standing, the _____ of a new organism must be deposited in a culture collection.

11. The type strain is the _____ strain for a defined species.

12. In Eukarya _____ are thought to have evolved from a symbiotic relationship with bacteria.

13. The science of identification and classification and naming of organisms is known as _____.

14. Phylogeny is the _____ of an organism.

15. _____ is a collection of different species, with each member sharing one or more major properties.

KEY WORDS AND PHRASES

Archaea groups (626)	Bacteria (p 625)
Bergey's Manual (p 629)	binomial system (p 629)
distance-matrix method (p 620)	domains (p 624)
Endosymbiotic theory (p 618)	evolutionary distance (p 620)
genus (p 609)	nomenclature (p 630)
phenotypic (characteristic) analyses (p 631)	phylogenetic trees (p 620)
RNA sequences (p 620)	signature sequences (p 609 and 623)
species (p 609)	stromatolites (p 610)
taxonomy (p 629)	

MULTIPLE CHOICE

1. All of the following gases are thought to have been prevalent in the atmosphere of the primitive earth EXCEPT:

    (A) ammonia   (D) nitrogen

    (B) carbon dioxide   (E) oxygen

    (C) methane

2. Place the evolutionary origin of these metabolic processes into the correct order, based on present knowledge.

   1. fermentative metabolism

   2. anaerobic respiration

   3. aerobic respiration

   4. anoxygenic photosynthesis

   5. oxygenic photosynthesis

   (A) 1,2,3,4,5 (B) 1,2,4,5,3 (C) 2,1,5,3,4 (D) 4,1,2,5,3

3. Mitochondria are descendants of

   (A) fermentative bacteria

   (B) cyanobacteria

   (C) Archaea

   (D) aerobic bacteria

   (E) none of the above

4. The molecule which has been most useful as an evolutionary chronometer has been

   (A) cytochrome c   (D) 16s RNA

   (B) DNA   (E) ferredoxin

   (C) 5s RNA

5. The ribosomes of which organisms are resistant to virtually all antibiotics?

   (A) eukaryotes

   (B) microsporidia

   (C) sulfur-dependent Archaea

   (D) viruses

   (E) thermotogans

6. If the DNA base compositions of two organisms are significantly different, then they

   (A) are unrelated

   (B) are unlikely to be related

   (C) may possibly be related

   (D) are likely to be related

7. Cell walls of Eukaryotes and Archaea microorganisms never contain:

   (A) chitin

   (B) cellulose

   (C) peptidoglycan

   (D) glycoprotein

8. The microbial 16s rRNA is similar to what eukaryotic component?

    (A) 16s rRNA

    (B) 32s rRNA

    (C) 18s rRNA

    (D) not equivalent to any eukaryotic system.

9. Methanogenesis is restricted to which group of organisms?

    (A) Bacteria

    (B) Archaea

    (C) Eukarya

    (D) Bacteria and Archaea

10. The percentage G+C ratio gives an indication of the makeup of a cells DNA. The ratio is taken from which one of the following equations.

    (A) $(G + C / A + T + G + C) * 100$

    (B) $(G + C / G + C) * 100$

    (C) $(G + C / A + T) * 100$

    (D) $(G + C / T + G + C) * 100$

10. In their membranes, Archaea have ether-linkages between glycerol and fatty acids. The Bacteria have:

    (A) ether-linkages

    (B) ether or ester linkages

    (C) ester linkages

    (D) neither ether or ester linkages

11. The inclusion of endosymbiotic cells into evolving Eukarya cell lines is the basis of what theory?

    (A) spontaneous insertion

    (B) endosymbiotic theory

    (C) Eukarya chronometers theory

    (D) Bergey's theory of evolution

13. Metabolism in primitive organisms is thought to have involved the use of $FeCO_3$ and H_2S. What condition would have been needed to allow these reactions to occur and what product would have formed.

    (A) anaerobic/pyrite

    (B) aerobic/pyrite

    (C) anaerobic/free Fe and S

    (D) aerobic/free Fe and S

14. Why is DNA thought to have evolved from RNA?

    (A) RNA had genetic and enzymatic capacities, but DNA is more stable

    (B) RNA lacked genetic and enzymatic capacities

    (C) RNA evolved from DNA

    (D) RNA and DNA evolved from proteins

DISCUSSION

1. Scientists believe that microbial life originated 3-3.5 billion years ago. What type of evidence do they have for this claim?

2. Why do evolutionary biologists think that the first organisms were heterotrophic and anaerobic?

3. What group of microorganisms were responsible for introduction of O_2 into the earth's atmosphere?

4. Cite three pieces of evidence that are consistent with the idea that mitochondria and chloroplasts of eukaryotic cells originated as endosymbiotic prokaryotic cells.

5. By what characteristics can Bacteria, Archaea, and Eukarya cells be distinguished?

6. What is meant by a bacterial species? How does the concept differ from its meaning in zoology and botany?

7. How do the principles of numerical taxonomy differ from those of conventional taxonomy?

8. If the DNA of two bacterial strains have the same G + C content, are the strains closely related? Why or why not?

9. What is 16S RNA? Give two reasons why analysis of this molecule has been very useful in determining the relatedness of microbial species.

10. List three morphological properties that are given heavy consideration in conventional classification schemes.

ANSWERS

Completion

1. radiodating; 2. metabolism, self-replication; 3. electron transport systems; 4. endosymbiont; 5. proteins, nucleic acids; 6. signature sequences; 7. Bacteria; 8. eukaryotes, Archaea; 9. species; 10. type strain; 11. reference; 12. mitochondria; 13. taxonomy; 14. evolutionary history; 15. genus

Multiple choice

1. E; 2. B; 3. D; 4. D; 5. C; 6. A; 7. C; 8. C; 9. B; 10. A; 11. C; 12. B; 13. A; 14. A.

Discussion

  1. See text Section 15.1

  2. See text Section 15.1

  3. See text Section 15.1

  4. See text Sections 15.5 and 15.6

  5. See text Section 15.8

6. See text Section 15.9
7. See text Section 15.9
8. See text Section 15.9
9. See text Sections 15.5 and 15.6
10. See text Section 15.9

Chapter 16
PROKARYOTIC DIVERSITY: BACTERIA

OVERVIEW

Chapter 16 (p 637-743) is divided into 32 sections, each of which describes the biology of a major physiological or morphological group of Bacteria. In this chapter, organisms with common phenotypic characteristics are grouped together. In Chapter 15, a grouping of microorganisms was presented on the basis of their evolutionary relationships.

This chapter contains a great deal of detailed and specific information about many bacterial genera. To a large degree, it can serve as a reference of detailed information about specific microbes. It should be referred to while learning about another aspect of microbiology such as ecology, pathogenesis, physiology, or industrial microbiology. In this summary, the key features of the groups are emphasized.

CHAPTER NOTES

Purple and green bacteria can produce **bacteriochlorophyll** pigments, and are capable of **photosynthesis**. They are distinguished from the cyanobacteria in that they do not evolve oxygen gas during photosynthesis. Photosynthetic growth only occurs under **anaerobic** conditions. All species can use reduced sulfur compounds as electron donors for photosynthesis. The green bacterium *Chloroflexus* and the non-sulfur purple bacteria can also use organic compounds as a source of reducing power under these conditions; these organisms are nutritionally versatile in that they can also grow aerobically as organotrophs. Green bacteria and purple bacteria differ in the types of bacteriochlorophyll they contain. In addition, the type of photosynthetic apparatus differs. Purple bacteria contain elaborate **internal membrane** systems, whereas green bacteria possess **chlorosomes** attached to the cytoplasmic membrane. Purple bacteria that grow autotrophically use the Calvin cycle to fix CO_2, whereas green bacteria use other reactions under autotrophic conditions.

Most species of **Cyanobacteria** are obligate photoautotrophs, which carry out **oxygenic photosynthesis** and fix CO_2 using the Calvin cycle. Their photosynthetic pigments are chlorophyll a and phycobilins. The cyanobacteria are morphologically diverse. Both unicellular and filamentous forms exist. In addition, some species produce differentiated cells. These include **akinetes**, which serve as resting stages, and **heterocysts**, specialized cells which fix N_2.

Prochlorophytes are organisms that contain the same pigments (chlorophyll a and b) and use the same photosynthetic mechanism (oxygenic photosynthesis) as plant chloroplasts. However, they are **prokaryotes** like cyanobacteria.

Chemolithotrophs can use a reduced inorganic compound as an energy source. If grown autotrophically, CO_2 is fixed via the Calvin cycle. However, many species can also grow as organotrophs. These organisms are subdivided on the basis of the inorganic compound that can be used for energy. **Nitrifying bacteria** really comprise two groups of bacteria -- one that oxidizes ammonia to nitrite, and a second that uses the nitrite to form nitrate. These organisms contain extensive internal membrane systems. **Sulfur-oxidizing bacteria** produce sulfuric acid from reduced sulfur compounds. Some species can grow at extremely low pH. **Chemolithotrophic hydrogen-oxidizing bacteria** are **facultative chemolithotrophs**; they also grow organotrophically. Many bacterial genera contain strains that grow chemolithotrophically on H_2.

They contain a membrane-bound **hydrogenase** that is associated with an electron-transport system that generates a proton-motive force.

Methylotrophs are oxidizers of one-carbon compounds and encompass a variety of organotrophic genera; they use compounds such as methanol, methylamine, or formate. **Methanotrophs** are more specialized; they do not grow on organic compounds, and are limited to substrates with just one carbon. They are also distinguishable from methylotrophs in that they grow on **methane gas**. They are aerobes, in part because the first enzyme involved in methane oxidation is an oxygenase. There are two groups of methanotrophs, which can be distinguished by their pathways of carbon assimilation and by the arrangement of internal membranes.

Sulfate- and sulfur-reducing bacteria are **obligate anaerobes** that use organic compounds as energy sources. Sulfate or elemental sulfur serves as the terminal electron acceptor in anaerobic respiration; hydrogen sulfide is the product. Two physiological groups are recognized; one oxidizes organic compounds to acetate, whereas the other oxidizes fatty acids completely to CO_2. These organisms are geochemically important, and are widespread in sulfate-rich environments where microbial decomposition of organic matter has created anaerobic conditions.

Homoacetogenic bacteria grow **anaerobically** either by using CO_2 as the terminal electron acceptor in anaerobic respiration or by fermenting sugars. In either case, the product is acetate. The **acetyl CoA pathway** is used to reduce CO_2 to acetate; this pathway is also used by methanogens and sulfate-reducing bacteria that are growing autotrophically.

Budding and Appendage (Prosthecate) bacteria are a group that contains bacteria that have many different types of cytoplasmic extrusions. These cellular appendages differ from fimbriae and pili in that they are extrusions that contain cytoplasm and are bounded by a cell wall. Diverse physiological types of bacteria contain prostheca. Prosthecate and budding bacteria provide interesting subjects for the study of cell differentiation, because the products of cell division are not identical, as in the case of binary fission. Included in this group are the **stalked bacteria** and the **budding bacteria**.

Spirilla are a diverse group of aerobic organotrophs, which are grouped on the basis of their morphology, includes organisms that are important in the decomposition of organic matter in the environment. It also includes representatives capable of **magnetotaxis** due to the presence of intracellular magnetite, and predatory bacteria that attack bacterial cells.

Spirochetes are noteworthy for their helical shape, and mechanism of motility. At each pole of the cell, **axial fibrils** are anchored; these structures are similar in chemical composition to flagella, but are wrapped around the cell. The rotation of the axial fibrils causes snakelike contortions of the cell which result in its movement.

Gliding bacteria do not have flagella, but can propel themselves when in contact with surfaces. The precise mechanisms of gliding motility have not been defined, but may involve rotating structures near the cell surface or the secretion of chemical surfactants. The **fruiting myxobacteria** have a complex life cycle, which culminates in the production of fruiting bodies containing resting structures called myxospores. The formation of these structures requires the cooperation of large numbers of cells. Aggregates of cells form through a chemotactic response.

Sheathed bacteria have a unique life cycle involving the formation of flagellated swarmer cells within a long tube or sheath. Under environmentally favorable conditions the cells remain in the sheath. When stressed, the cells leave the sheath looking for more favorable conditions. Three genera are recognized.

Pseudomonas is an important genus of Gram-negative rods with polar flagella. These aerobic organotrophs do not carry out fermentative metabolism. As a group, these organisms can utilize a wide variety of organic compounds. However, they are unable to break down polymers.

153

Free-Living aerobic nitrogen-fixing bacteria are capable of fixing N_2 gas aerobically. They are, for the most part, soil bacteria and play a significant role as a source of reduced N in many natural ecosystems.

Acetic acid bacteria are industrially important because they can spoil alcoholic beverages by oxidizing ethanol to acetic acid under aerobic conditions. They are used to produce vinegar (dilute acetic acid) in industrial processes.

Vibrios are aquatic Gram-negative bacteria that are similar to enteric bacteria in that they are facultative aerobes that use fermentations under anaerobic conditions. They differ from enteric bacteria in that they are oxidase positive, and from pseudomonads in their ability to grow anaerobically. This group includes the **luminescent** bacteria. These are often associated with special light organs in fish. Luminescence is produced via the enzyme **luciferase**; in this reaction reducing power is consumed as electrons are transferred to O_2.

Facultatively aerobic Gram-negative bacteria are characterized as Gram-negative rods that are facultative aerobes and oxidase negative. Under anaerobic conditions, they ferment sugars. The group can be subdivided by the pattern of fermentation products -- the amount of organic acids and the presence of butanediol are the distinguishing characteristics. Many strains in this group are **pathogens** for humans, animals, or plants. *Salmonella*, *Shigella*, and *Yersinia* species are generally human or animal pathogens. Some species of *Erwinia* are plant pathogens.

Neisseria and Gram-negative cocci are a diverse group of bacteria that are related by cell wall characteristics, morphology, a lack of motility, nonfermentative aerobic metabolism and a similar DNA base composition. The major distinction within the group is based on the oxidase reaction.

Rickettsia are **intracellular parasites** unable to live outside an animal host cell. The cells have a normal Bacterial morphology. However, their metabolic capacity is limited. They can generate energy by electron-transport phosphorylation and synthesize some of the monomers found in macromolecules. Their membranes appear leakier than normal, and they may assimilate some coenzymes from their hosts. These pathogens are transmitted by arthropod vectors.

Chlamydia are pathogens that are generally transmitted through the air. These cells have fewer metabolic functions than rickettsias. They cannot generate ATP themselves, and must obtain it from the host cell.

Gram-positive cocci contains members of widely different physiological characteristics. Genera are distinguished by the arrangement of cells and the capacity for fermentative growth. These bacteria are rather resistant to drying, and therefore survive well in air or on the surface of the skin.

Lactic acid bacteria produce lactic acid as the major or sole product of sugar fermentation. Other shared characteristics are a limited capacity for biosynthesis of amino acids and other macromolecular precursors, and the absence of electron transport phosphorylation. Thus, even in the presence of oxygen, these bacteria have a fermentative metabolism. The group is subdivided by the pathway of sugar fermentation. **Homofermentative** species produce lactic acid as the sole product. In **heterofermentative** species, ethanol and carbon dioxide are produced in addition to lactic acid.

Many species in this group are important in the production of fermented foods, especially dairy products. β-**hemolytic** streptococci often cause strep throat in humans.

Endospore-forming bacteria are distinguished by their morphology, relationship to O_2, and energy metabolism. The shape and cellular location of endospores are of taxonomic value in distinguishing species within a genus. These organisms are genetically heterogeneous, but ecologically related. All occur primarily in soil; the endospore provides a survival mechanism in the highly variable soil environment. *Bacillus* strains often produce extracellular hydrolases which attack macromolecules. **Antibiotics** may be produced during sporulation. *Clostridium* species are obligate anaerobes that can only synthesize ATP by substrate-level phosphorylation.

Mycoplasmas are bacteria that lack cell walls. However, the nature of their cell membrane confers resistance to osmotic lysis. The membranes contain **sterols** or other compounds that enhance their stability. Some mycoplasma cells are only 0.2 micrometers in diameter; these are the smallest free-living cells known.

Actinomycetes encompass a vast variety of bacteria most of which have GC ratios between 60 and 70 mole percent. All of the actinomycetes are Gram-positive, rod shaped and nonmotile. Eight groups of actinomycetes are recognized. These include *Streptomyces* which produce a wide number of antibiotics.

Coryneform bacteria are Gram-positive aerobic bacteria that occur in irregularly-shaped arrangements. They are widely distributed, and one genus **Arthrobacter** is commonly found in soil.

Propionic acid bacteria produce propionic acid. They are encountered in Swiss cheese where the fermentation of lactate releases CO_2 that causes the holes, and the formation of propionic acid provides some of the flavor.

Mycobacteria are rod-shaped bacteria that have distinctive staining properties. They are pleomorphic and undergo filamentous growth. The virulence of *M. tuberculosis* is in part related to the filamentous structure. However, a true mycelium is not formed as slight disturbance of the mycelium will shatter it into its rod or coccoid elements. A characteristic of the group is their ability to form yellow carotenoid pigments.

Filamentous actinomycetes produces a network of branching filaments, called a **mycelium**, which is analogous to that formed by filamentous fungi. Growth occurs at the tips of the filaments. When cultured on agar medium, aerial filaments are formed which give rise to **conidia**. The shape and arrangement of these aerial structures are used in taxonomy. These bacteria are found in soil, where they are responsible for the typical odor of soil. In addition, many produce antibiotics, some of which are medically important.

Chapter 17
PROKARYOTIC DIVERSITY: ARCHEAE

OVERVIEW

Chapter 17 (p 744-788) focuses on the members of the domain Archaea. While the Archaea are prokaryotes, they are as distinctive from the Bacteria as they are from the eukaryotes. The Archaea were once known as archaebacteria and live in many extreme environments. The Archaea can be divided into four groups: the methanogens, the halophiles, the hyperthermophiles and the genus *Thermoplasma*. The distinctive features of this group were also briefly discussed in Section 18.7 and this section should be reviewed

CHAPTER NOTES

The Archaeal membranes differ from Bacterial membranes in that they contain **ether-linked lipids** bonded to glycerol. The ether-linked lipids are common to all Archaea. Glycerol diethers and diglycerol tetraethers are the major types of lipids present in the cell membrane. The Archaea also contain large amounts of non-polar lipids. The overall arrangement of the cell membrane is similar to that found in Bacteria and Eukarya. The Archaea can alter the thickness of their membrane by including or removing pentacyclic rings in the structure.

Archaeal cell walls do not contain muramic acid and D-ammino acids, the building blocks of peptidoglycan; particular species may contain **pseudopeptidoglycan**, polysaccharide, **glycoprotein**, or protein in their cell walls.

Metabolism in the Archaea is varied, ranging from chemoorganotrophic reactions to autotrophic utilization of CO_2. In general, the types of metabolism within the group are similar to what is found in the Bacteria. The major exception are the reactions leading to methanogensis.

The six genera of extremely halophilic Archaea inhabit **hypersaline** environments and will not grow at NaCl concentrations less than 1.5 molar. All extreme halophiles can grow at salt concentrations near the salt's saturation point. Two genera are not only halophilic, but also **alkalinophilic**. That is, they grow best at pH values above 9. These organotrophic bacteria require Na^+ ions to stabilize their glycoprotein cell wall. The high external salt concentration is balanced by the intracellular accumulation of K^+ ions.

Certain species of *Halobacterium* can synthesize ATP using light energy. However, the process does not involve chlorophyll pigments as in photosynthesis, but rather a membrane protein called **bacteriorhodopsin**. The absorption of light by **retinal** associated with this protein is used to pump protons across the cell membrane. The resulting proton motive force can drive ATP synthesis via a membrane-bound ATPase.

Methanogens are strict anaerobes which convert one of three classes of substrate: CO_2, methyl compounds, or acetate to methane gas. The formation of methane can be viewed as a type of anaerobic respiration. These organisms contain a unique set of coenzymes which are necessary for the reduction of C-1 intermediates to methane; for example, **coenzyme M** is involved in the final step of methane formation. When CO_2 is the carbon source, the **acetyl CoA pathway** is used to produce organic carbon, rather than the Calvin cycle. Although their physiological diversity is limited, methanogens comprise seven morphological groups.

Hyperthermophilic Archaea includes the most thermophilic of all known prokaryotes. All representatives require reduced sulfur compounds for their metabolism; in most cases, reduced sulfur is used as an **electron acceptor** to carry out anaerobic respiration. However, *Sulfolobus* can grow autotrophically using elemental sulfur as an energy source.

Thermoplasma is a cell-wall-less prokaryote similar to the mycoplasmas. However, it is an acidophilic, aerobic chemoorganotroph that is also thermophilic. It is generally found in self-heating coal refuse piles. The cell membranes of the organism are chemically unique, containing lipopolysaccharide that consists of tetraether lipid with mannose and glucose subunits.

EUKARYA: EUKARYOTIC MICROORGANISMS

OVERVIEW

Chapter 18 (pages 773-788) focuses on the eukaryotic microbes: algae, fungi, slime mold and protozoa. It should be clear that eukaryotic microorganisms differ from Bacteria and Archaea in many ways. These differences include cell size, internal structure, genetic arrangement and evolutionary history.

CHAPTER NOTES

Algae contain chlorophyll, and carry out oxygenic photosynthesis with water serving as the electron donor. Specific groups are distinguished by the type of accessory pigment they contain, the chemical composition of carbon reserve compounds and cell walls, and motility. Algae can be either green, containing only chlorophyll, or brown and red. The later contain not only chlorophyll but pigments such as carotenoids that hide the green color. The chlorophyll is contained in membrane structures called **chloroplasts**. Algae can occur as either single cells or as aggregates of cells. Algae are found primarily in aquatic habitats and at the soil surface. Algae should not be confused with cyanobacteria, which are bacteria.

Fungi are chemoorganotrophs, lack chlorophyll, and have simple nutritional requirements as compared to bacteria. Fungi can be differentiated from prokaryotes because they are much larger, contain a nucleus, vacuoles, and mitochondria. The three important groups of fungi are molds, yeasts, and mushrooms. Fungi are particularly important in the decomposition of wood or wood products such as paper. Unlike the Bacteria and Archaea, there is great diversity in both fungal morphology and sexual life cycles.

Molds are filamentous fungi that have widespread occurrence in nature. The morphology of molds is similar to that of the prokaryotic actinomycetes -- a surface **mycelium** and aerial **hyphae** that contain asexual spores (**conidia**).

Yeasts are unicellular fungi usually occurring as spheres, ovals or cylinders. They tend to favor environments rich in sugars, such as plant surfaces. However, they are also the causative agents of a number of important diseases. Asexual division in **yeasts** involves **budding**.

Mushrooms (Basidiomycetes) are filamentous fungi that form large above-ground **fruiting bodies**, although the major portion of the biomass consists of hyphae below ground.

Slime molds are non phototrophic eukaryotic microorganisms that live on decaying plant matter by phagocytizing microorganisms present on the surfaces. There are two groups of slime molds: **cellular** slime molds such as *Dictyostelium* that undergo a life cycle in which the cells exist independently as single amoebalike cells and **acellular** slime molds where the vegetative forms are naked masses of protoplasm called plasmodia.

Protozoa are unicellular microorganisms that lack cell walls and obtain nutrients by ingesting other microbes, or by ingesting macromolecules in solution in a process of **pinocytosis**. They lack pigments and may be motile. There are four major groups, which are distinguished by their mechanism of motility (flagella vs. pseudopodia vs. cilia vs. non-motile), as well as the characteristics of their life cycles. **Mastigophora** (flagellates) are motile through the use of flagella, **Sarcodina** (amoebas) are motile with amoeboid movement, **Ciliophora** use cilia for movement and **Sporozoa** are non-motile. Each group contains representatives that cause important human diseases.

HOST-PARASITE RELATIONSHIPS

OVERVIEW

In Chapter 19, pages (789-817) the principles of how microorganisms grow in the animal body as well as some of the defense mechanisms used by animals to resist infections are explained. Some microbial species can cause **infectious diseases**. The relationship between **pathogens** and their hosts is a dynamic one, in which the pathogen possesses attributes that promote **infection**, and the host has systems that kill and eliminate infecting microbes. The outcome of the confrontation between the pathogen's virulence factors and the host's defense mechanisms determines whether an infection occurs.

CHAPTER NOTES

Animals need defense mechanisms because they consist of rich nutrient sources for microbial growth, microbes use **virulence factors** to evade these mechanisms. Different anatomical parts vary in the type of microbes that are most likely to grow there, because these parts differ chemically and physically. Microbes occur in those body parts exposed to the outside world; note that this includes the gastrointestinal tract. Internal organs and the circulatory systems do not have a normal microbial flora.

The skin presents a formidable barrier to the penetration of microbes into the body. The sweat glands, which are moist, do contain a normal flora. The secretions of **sebaceous glands** are nutrient-rich and support a resident flora. Microbial **transients** that are deposited on the skin generally cannot grow because of the skin's low moisture content and acid pH.

The teeth provide abundant surface area for the attachment of bacteria. Two important characteristics of the oral microflora are their ability to attach to surfaces via glycocalyx, and the localized changes in oxygen and pH they create by their metabolism. The bacteria synthesize a sticky polysaccharide to adhere strongly to these surfaces -- the film of bacteria and polysaccharide is called **dental plaque**. Many anaerobic bacteria proliferate in the crevices near teeth and gum. Oxygen is depleted by facultative bacteria in the plaque, and by their action anaerobic **microenvironments** are created. Lactic

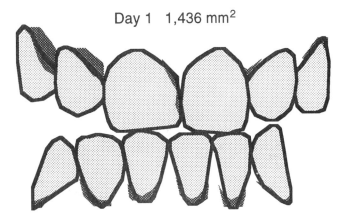

Day 1   1,436 mm^2

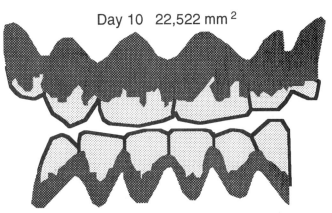

Day 10   22,522 mm^2

Distribution of dental plaque on brushed and unbrushed teeth.

acid bacteria produce organic acids, which lower the pH near the tooth surface and dissolve the calcium phosphate that makes up the tooth. This is the cause of tooth decay. This disease, called

dental caries, is caused by two streptococci, *S. sobrinus* and *S. mutans*. Tooth decay is promoted by diets high in **sucrose**, because *S. mutans* produces a polysaccharide adhesive specifically from sucrose, and the sugar is used as a substrate to produce organic acids which lower pH in the microenvironment near the tooth surface.

The bacterial flora of the **gastrointestinal tract** increases in magnitude as one travels down it. The low pH of the stomach kills many ingested bacteria, and prevents them from reaching the rest of the system. High numbers of bacteria are found in the **large intestine**. This is an anaerobic environment -- any oxygen present is quickly consumed by facultative bacteria. This region is the normal habitat of *Escherichia coli*, but it comprises far less than 1% of the total microbial flora.

The normal gastrointestinal flora varies with an individual's diet, and differs among different animal species. The microbial flora has metabolic effects on the animal; they may serve as sources of essential vitamins or transform animal metabolites such as **bile acids**.

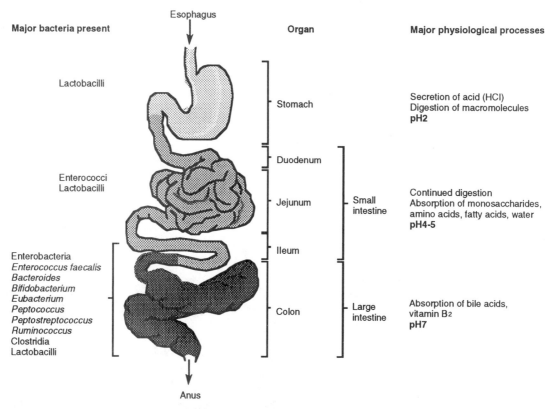

Bacteria proliferate in the **mucous membranes** found in the **upper respiratory tract**. Some bacteria that are pathogenic when they infect other parts of the body are parts of the normal flora in the **nasopharynx**. Most bacteria inhaled from the air are trapped in the nasal passages. The **lower respiratory tract** is kept sterile by this mechanism coupled with the action of cilia on the tissue walls. These cilia push particles upwards, away from the lungs.

Some host defense mechanisms are **nonspecific** in that they protect against a wide variety of pathogens. In many cases, the susceptibilities of animal species to particular pathogens are consequences of the animal's physiology, nutrition, anatomy, and types of receptors on tissues. This occurs because a pathogen may have rather specific requirements for entering into or growing in the host, and these are not met in all animal species.

An individual's susceptibility to infection increases as the functioning of defense mechanisms declines. An animal's defense mechanism is affected by factors such as age, stress, and diet.

A major nonspecific host defense is the anatomy of the host. Bacteria cannot penetrate intact skin. Furthermore, sebaceous glands secrete fatty acids that inhibit many microbes. Cilia and hair in the nasopharynx trap bacteria that are inhaled. The acidity of the stomach kills many of the bacteria that are ingested. Bactericidal substances are secreted in several parts of the body to prevent microbial growth.

However, these protection mechanisms are not absolutely effective. Some pathogens may evade them, especially in **compromised hosts**. These are individuals with weakened defense mechanisms. Low resistance may result from genetic constitution, living habits (smoking, drug usage, poor nutrition), or from disease. Hospital patients often have lowered resistance to infections because of the treatments to which they are subjected.

To begin an infection, a pathogen must enter the host. Many pathogens enter through mucous membranes (epithelial cells coated with glycoprotein); these interactions are specific in that pathogens have surface macromolecules that bind only to cell surface components of the specific tissue that becomes infected. These bacteria may then **adhere** strongly to these mucous membranes, because they synthesize a sticky **glycocalyx**.

To cause a disease, a pathogen that enters the host must be able to grow. This entails the bacterium obtaining the nutrients it needs for growth from the host. Animals have developed mechanisms to keep the trace element iron in short supply. Serum contains the protein **transferrin** that binds iron, and thereby makes it unavailable to microbes. Some bacteria counteract this mechanism by producing their own iron-chelating compounds, from which they can obtain iron.

Some virulence factors that promote infection are enzymes that help pathogens spread through the body. These include **hyaluronidase** and **collagenase** that break down substances that hold tissues together. Other enzymes dissolve **clots** by which the body walls off infections to keep microbes from spreading. In contrast, **staphylococci**, which cause boils and pimples, produce an enzyme that causes clots and thus they keep themselves insulated from host defense mechanisms.

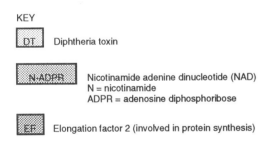

KEY

DT   Diphtheria toxin

N-ADPR   Nicotinamide adenine dinucleotide (NAD)
N = nicotinamide
ADPR = adenosine diphosphoribose

EF   Elongation factor 2 (involved in protein synthesis)

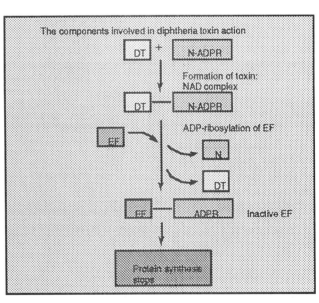

Not all pathogens need to spread from the site of entry to produce a debilitating disease. If they produce extracellular proteins called **exotoxins**, host damage can occur at sites far removed from a localized focus of infection. **Diphtheria** toxin is a good example. *Corynebacterium diphtheriae* colonizes the nasopharynx, and releases its toxin. Diphtheria toxin acts intracellularly upon eukaryotic cells to inhibit protein synthesis. The activity of the protein can be separated into two parts: fragment B is responsible for recognizing specific receptors on animal cells, and promoting the entry of fragment A into the cytoplasm. Fragment A catalyzes a reaction between NAD and an **elongation factor** involved in protein synthesis that inactivates the latter by adding **ADP-ribose** to it. The genetic information for toxin production is contained within a prophage carried by pathogenic strains of *C. diphtheriae*. Toxin production is stimulated when the iron concentration is low.

Clostridia are typically soil inhabitants, but some produce exotoxins. *C. tetani* can be introduced into the body with deep puncture wounds. If the wound becomes anaerobic, the organism can grow and release its toxin. *C. botulinum* does not grow in animals. However, it can grow in improperly processed food. Here it elaborates its exotoxin, and if the food is not cooked sufficiently to denature the protein, the ingested exotoxin can act.

Both toxins act on the nervous system, and are lethal if not treated properly. Tetanus toxin causes **spastic paralysis** by binding to a lipid in nerve synapses and blocking the action of inhibitory motor neurons. Botulinum toxin blocks the release of acetylcholine at nerve-muscle junctions. This prevents muscle contraction, resulting in **flaccid paralysis**. Some types of botulinum toxins are coded by lysogenic bacteriophage.

Exotoxins that acts in the small intestine are called **enterotoxins**. These cause **diarrhea**, the secretion of fluid into the intestinal passage. The pathogens enter the body in contaminated food or water. In the case of *Staphylococcus aureus*, the enterotoxin is produced by the growth of bacteria in food. In other cases, the bacteria colonize the small intestine and produce enterotoxin there. In some bacteria, the enterotoxin gene is located on a plasmid.

Vibrio cholerae causes a life-threatening diarrhea, and therefore the action of its toxin has been studied most intensively. As in diphtheria toxin, several functional parts of the protein can be recognized. These include the B subunit that recognizes an animal cell receptor, and the A_1 subunit that produces the toxic effect. It activates the enzyme **adenyl cyclase**, which synthesizes **cyclic AMP** in the animal cell. The cAMP signals the cell to excrete ions. The change in ionic balance causes water to leak from the cells into the intestinal lumen. Humans may lose ten liters of fluid per day. Treatment entails giving similar volumes of electrolyte solution orally.

All Gram negative bacteria contain **endotoxin**. It is the lipid portion of the lipopolysaccharide of the outer membrane. This material can induce fever in animals. Large doses cause a general inflammatory state, and can lead to death due to hemorrhagic shock. However, in general, endotoxin is much less toxic than exotoxins. A sensitive assay for the presence of Gram negative bacteria in materials is the reaction of endotoxin with cell lysates from the horseshoe crab.

Pathogens can cause severe disease in one of two ways: (1) by elaborating an exotoxin that produces effects far from the site of infection, or (2) by being highly **invasive** and growing profusely in a body tissue. Specific pathogens differ in the degree of toxigenicity and invasiveness. Pathogens maintained in laboratory culture may become **attenuated** in virulence, because **virulence factors**, the characteristics that are essential for infecting

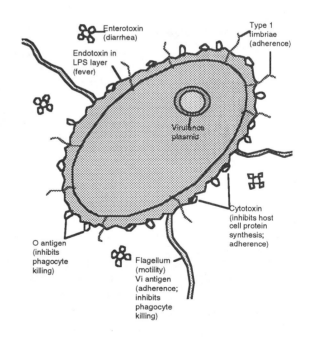

162

animals are nonessential in culture, and mutants that lack these will probably grow faster. Pathogens often contain a set of virulence factors, which act in concert to initiate an infection. These factors may be a toxin, responsible for adherence to host tissues, or enhance nutrient acquisition (see Figure 11.21).

Inflammation is a body response to injury that can be painful, but it is also a defense mechanism. The response increases the flow of circulatory fluid to the affected area, and brings phagocytic cells to it. In addition, a fibrin clot is formed to prevent microbes from reaching other parts of the body. Fever is induced by bacterial endotoxin, or by **endogenous pyrogens** released by phagocytes when they are active. Slight fevers enhance defense mechanisms by speeding phagocytosis and antibody reactions.

SELF TESTS

COMPLETION

1. Infections occur when pathogens _____ the host.

2. Host-pathogen interactions are dynamic. The host resists infection with a variety of _____ and the pathogen has _____ to evade these mechanisms.

3. Bacteria on the skin normally grow near _____.

4. Dental caries requires a diet including _____ and the bacteria _____ and _____.

5. Most bacteria in the gastrointestinal tract grow best when oxygen is _____.

6. Potentially pathogenic bacteria may colonize the _____ respiratory tract.

7. Colonization of the intestinal tract by pathogens is made more difficult by the presence of _____.

8. The most effective barrier to penetration of microorganisms into body tissues is the _____.

9. Iron in the body is rendered unavailable to invading pathogens by binding to _____.

10. _____ are bacterial enzymes that degrade animal cell membranes. Their presence is assayed by exposing _____ to the bacteria.

11. Several exotoxins can be separated into fragments, one of which is responsible for the toxic effect on animal cells, whereas the other is necessary for _____.

12. Cholera toxin stimulates the activity of _____ in animal cells.

13. The amount of endotoxin required to kill an animal is _____ than the lethal dose of tetanus toxin.

14. When pathogenic bacteria are maintained in laboratory culture, their virulence often _____.

15. A slight fever enhances body resistance to infection by accelerating the activity of _____.

KEY WORDS AND PHRASES

bacteremia (803)	blood agar plate (p 803)
colonization (p 802)	colonization factor antigens (p 801)

dental plaque (p 794)	dental plaque (p 794)
disease (791)	endotoxin (p 809)
enterotoxin (p 807)	exotoxin (p 805)
epithelial cells (p 791)	flatus (p 799)
focus of infection (p 803)	food poisoning (p 807)
Fusobacterium (p 794)	hemolysins (p 803)
host (p 790)	infection (791)
inflammatory response (p 814)	invasion (p 802)
large intestine (p 796)	leukocytes (p 814)
leukocidins (p 804)	*Limulus* assay (p 809)
localized infection (803)	lockjaw (p 806)
mucous membrane (p 791	normal flora (p 793)
opportunistic pathogens (p 798)	parasite (p 790)
pathogen (p 790)	pathogenicity (p 790)
Propionibacterium (p 793)	pyogenic (p 814)
pyrogens (p 809)	resident microbes (p 793)
small intestine (p 796)	sebaceous gland (p 792)
specific adherence (p 800)	*Staphylococcus* (p 793)
systemic infection (p 803)	transient microbes (p 793)
virulence (p 791)	virulence factor (p 810)

MULTIPLE CHOICE

1. Which of the following is **not** a virulence factor of a pathogenic microbe?

    (A) induction of fever in an infected host

    (B) resistance to phagocytosis

    (C) adherence to mucous membranes

    (D) production of hyaluronidase

    (E) possession of a capsule

2. Which of the following are exotoxins?

    1. the toxin responsible for botulism

    2. the lipopolysaccharide of *Salmonella typhi*

    3. *Staphylococcus aureus* enterotoxin

    4. lysozyme

    (A) 1,2 (B) 1,3 (C) 2,4 (D) 3,4 (E) 1,4

164

3. Adherence to mucous membranes of the host is an important virulence factor in many pathogenic bacteria. The ability to adhere may be due to:

    (A) cell wall   (D) B and C

    (B) glycocalyx   (E) all of the above

    (C) pili

4. A capsule increases the virulence of certain pathogenic bacteria by:

    (A) aiding the cell to evade phagocytosis

    (B) destroying leukocytes

    (C) inhibiting the formation of interferon

    (D) destroying red blood cells

    (E) none of the above

5. The endotoxin of Gram-negative bacteria is:

    (A) only found in lysogenic strains

    (B) protein

    (C) bound to the ribosomes

    (D) the lipopolysaccharide of the cell wall

    (E) the lipid of the cell membrane

6. The direct cause of the symptoms of tetanus is:

    (A) endotoxin

    (B) exotoxin

    (C) toxic effect of the cells

    (D) hemolysin

    (E) collagenase

7. All of the following are useful in preventing infectious disease EXCEPT:

    (A) mucous membranes   (D) inflammation

    (B) epithelial cells   (E) fever

    (C) alcoholic beverages

8. The occurrence of an infectious disease depends upon

    (A) the resistance of the host

    (B) the virulence of the pathogen

    (C) infection of the host by the pathogen

    (D) A and B above

    (E) all of the above

9. Each of the following enhances the pathogenicity of a microbe EXCEPT:

    (A) ability to grow intracellularly in phagocytic cells

    (B) possession of a capsule

    (C) ability to alter surface composition

    (D) exotoxin production

    (E) production of pyrogenic compounds

10. Exotoxins have all of the following properties EXCEPT:

    (A) they act extracellularly from the microbe that produced them

    (B) they are proteins

    (C) they induce fever

    (D) the information for exotoxin production may reside on a prophage

    (E) they can damage tissue at sites removed from the area of infection

11. Which of the following is not a host resistance factor?

    (A) lipopolysaccharide cell wall layer

    (B) skin

    (C) inflammation

    (D) macrophages

    (E) interferon

12. A fever is usually one host response to bacterial invasion. This response is due to:

    (A) a potent toxin released by the bacterium

    (B) the lipopolysaccharide of the bacterial cell wall

    (C) histamine released by bacteria

    (D) red blood cell lysis

    (E) loss of the normal microbial flora

13. Which of the following areas of the human body is usually colonized by bacteria?

    (A) central nervous system   (D) lungs

    (B) urinary tract   (E) stomach

    (C) large intestine

MATCHING

I. Match the bacterial group with the portion of the gastrointestinal tract where they reside.

Bacterial Group	Location
1. Lactobacilli	(A) colon
2. Enterococci	(B) ileum
3. Enterobacteria	(C) jejunum
4. Clostridia	(D) duodenum
5. *Peptococcus*	(E) stomach

II. Match the exotoxin producing organism with the disease.

Organism	**Exotoxin**
1. *Clostridium botulinum*	(A) tetanus
2. *Clostridium perfringens*	(B) botulism
3. *Corynebacterium diphtheriae*	(C) cholera
4. *Clostridium tetani*	(D) pyogenic infections: tonsillitis
5. *Escherichia coli*	(E) gas gangrene
6. *Yersinia pestis*	(F) plague
7. *Bordetella pertussis*	(G) toxic shock, boils, scalded skin
8. *Streptococcus pyogenes*	(H) gastroenteritis
9. *Vibrio cholerae*	(I) diphtheria
10. *Shigella dysenteriae*	(J) whooping cough
11. *Staphylococcus aureus*	(K) dysentery

DISCUSSION

1. Distinguish between colonization, infection, and disease.

2. What characteristics of the skin make it a difficult site for bacteria to grow? In what regions of the skin can bacteria proliferate?

3. Discuss the conditions that are necessary for tooth decay to occur What type of bacterial activities are essential for this process?

4. What mechanisms keep the lungs a microbially sterile environment?

5. Cite three pieces of evidence that the normal bacterial flora are beneficial to the host in terms of nutrition and defense against pathogenic bacteria.

6. What environmental factors can make an individual more susceptible to infection?

7. What role does inflammation have in enhancing the resistance of host?

8. What factors determine whether an *E. coli* strain will be an intestinal pathogen?

9. Describe the mechanism of action of four enzymes that act as virulence factors for bacteria.

10. Contrast the chemical composition and mechanism of action of endotoxins and exotoxins.

11. How does cholera toxin cause diarrhea?

12. Prophage repressor protein (see Section 6.12) and interferon can protect cells from virus infection. Compare the stimuli for production of these substances, the range of viruses against which protection is effective, and the mechanism of protection.

ANSWERS

Completion

1. grow in; 2. defense mechanisms, virulence factors; 3. sweat glands; 4. sucrose, *Streptococcus mutans*, *S. sobrinus*; 5. absent; 6. upper; 7. the normal microbial flora; 8. skin; 9. transferrin; 10. hemolysins, red blood cells; 11. binding to host cells; 12. adenyl cyclase; 13. greater; 14. decreases; 15. phagocytic cells.

Multiple choice

1. A; 2. B; 3. D; 4. A; 5. D; 6. B; 7. C; 8. E; 9. E; 10. C; 11. A; 12. B; 13. C.

Matching

I. 1. A,B,C,D,E; 2. C; 3. A,B; 4. A, B; 5. A, B.

II. 1. B; 2. E; 3. I; 4. A; 5. H; 6. F; 7. J; 8. D; 9. C; 10. K.

Discussion

1. See introduction and Sections 19.7 and 19.8

2. See text Section 19.2

3. See text Section 19.3

4. See text Section 19.5

5. See text Section 19.4

6. See text Section 19.7

7. See text Sections 19.13

8. See text Section 19.8

9. See text Section 19.9

10. See text Section 19.5

11. See text Section 19.4

12. See Sections 19.12 and 19.13

Chapter 20
CONCEPTS OF IMMUNOLOGY

OVERVIEW

Chapter 20 (pages 817-868) is concerned with the principles of how the immune system functions in vertebrate animals. The immune response is a sophisticated system to neutralize and remove foreign substances from the animal's body. It can be highly specific, it has a memory so that faster responses occur after the first experience with a foreign material, and it can distinguish between foreign and "self" molecules.

CHAPTER NOTES

The definitions for **antigen** and **antibody** are circular -- they are defined in terms of the other. An antigen, or more specifically an **immunogen**, induces an immune response which can include the formation of antibodies (serum proteins), the activation of T cells or both. The antibodies can bind to the antigen. Only large molecules induce an immune response; proteins are most effective in doing so, but other macromolecules are also immunogenic. Despite the requirement for a large size, a specific antibody reacts with only a small

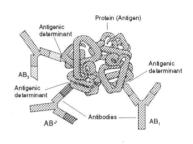

portion of the immunogenic molecule. This **antigenic determinant** of an immunogen may comprise only 4-5 amino acids. Thus, a protein may have a large number of different antigenic determinants, so that the immune response to it can consist of a number of antibodies with different specificities.

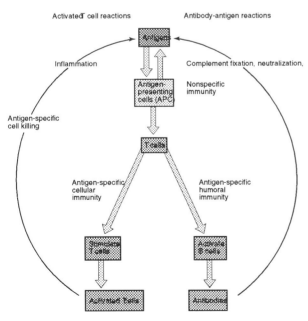

Lymphocytes and **macrophages** are the two most important types of cells in the immune system. Both arise from **stem cells** in the bone marrow. There are two types of lymphocytes: **B cells** and **T cells**. They are distinguished by the organ in which they differentiate (bone marrow versus thymus), and the differentiation leads them to have distinct functions. There are several functional types of T cells. Macrophages are found fixed to tissue surfaces in the spleen and lymphoid tissue. Here they phagocytize foreign particles, and by **processing** these antigens and **presenting** them to lymphocytes are essential to developing the immune response.

Antibody proteins in serum are found in the gamma globulin fraction. There are five distinct classes of **immunoglobulins**. **IgG** is the most prevalent, and comprises 80% of total immunoglobulins. Antibodies from several different classes are made in response to antigens.

An IgG molecule consists of four polypeptide chains, linked by disulfide bridges. The four polypeptides consist of two identical **light chains** and two identical **heavy chains**. Each Y-shaped molecule contains two binding sites for antigen; these are located at the ends of the "Y". The amino terminal portions of both the light and heavy chains are involved in antigen binding. The interaction between antigen and binding site is analogous to what occurs between a substrate and the active site of an enzyme. Both the light and heavy chains can be subdivided into **constant** regions in which the amino acid sequences are identical in all IgG molecules, and **variable** regions which are unique to the specific antibody. The variable regions comprise the antigen-binding site, and it is variation in the amino acid sequence of this portion which results in different three dimensional shapes to produce antibodies of unique specificity.

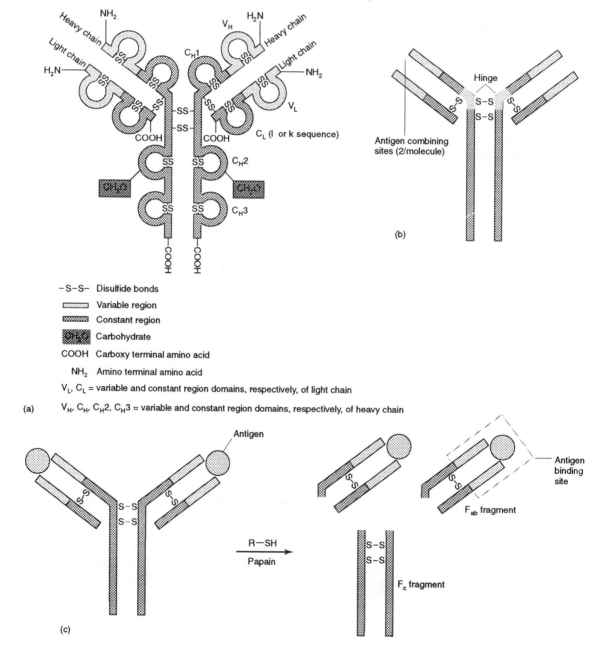

-S-S- Disulfide bonds

▭ Variable region

▨ Constant region

▨ Carbohydrate

COOH Carboxy terminal amino acid

NH₂ Amino terminal amino acid

V$_L$, C$_L$ = variable and constant region domains, respectively, of light chain

(a)   V$_H$, C$_H$, C$_H$2, C$_H$3 = variable and constant region domains, respectively, of heavy chain

170

How do the different classes of antibodies differ? Each of the five classes (see text Table 12.2) has a unique sequence in the constant region of the heavy chain which is found in all antibodies of that class. The other classes also differ in structure and function from IgG. **IgM** is an aggregate of five antibody molecules. Thus, it has ten antigen combining sites. It comprises 10% of serum antibody and is typically the first class of immunoglobulin made in response to an immunogen. **IgA** is present in secretions. **IgE** is bound to mast cells and basophils, and is responsible for initiating allergic reactions induced by those cells.

Let us now consider how **antibodies are formed** in response to a foreign substance. A related problem is how an animal has the capacity to quickly respond to any of the foreign materials present in the environment -- molecules it has never encountered previously.

Antibody production requires the interaction of B cells, T-helper (CD4) cells, and macrophages. Macrophages phagocytize foreign particles such as bacteria, digest them, and portions become attached to the macrophage cell surface. In this way, the macrophage can **present** the antigen to B and T lymphocytes. Both the B and T cell have synthesized surface receptors complementary to the antigen on the macrophage surface. In other words, the B lymphocyte is already genetically programmed to produce the antibody which binds to this antigen. The presentation by the macrophage serves to select the correct B cell out of the large population of B lymphocytes. The role of the CD4 cell is to secrete **helper factors** which stimulate differentiation and multiplication of the B cell. T cell receptors recognize the foreign antigen, as well as MHC class II antigens. After stimulation, the B cell forms a clone of identical B cells, which differentiate into **plasma cells** that secrete large amounts of the specific antibody.

This production of plasma cells that follows the animal's first exposure to an antigen is called the **primary antibody response**. However, these cells are short-lived, and the antibody **titer** decreases within a few months. But long-lived **memory cells** also arise from the clone of specific B cells, and if the animal is challenged with the specific antigen a second time, these cells quickly transform to plasma cells which secrete antibody. This **secondary response** occurs more rapidly and produces higher antibody titers than the primary response.

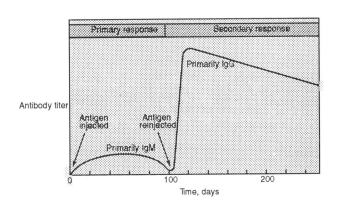

Many immune reactions involve the recognition of antigens on animal cell surfaces. Especially important are the **major histocompatibility complex** (MHC) antigens. Class I and Class II MHC antigens are embedded in cell membranes; they serve as identifying markers for T lymphocytes to recognize "self" cells. In addition, these molecules have antigen binding sites, which "present" antigens to T_c or T_h cells which bind to specific MHC antigen classes.

There are several types of T lymphocytes, which can be distinguished by the type of antigen receptors on their surface. These proteinaceous receptors are analogous to antibodies in that the proteins contain constant and variable regions, and bind specifically to particular molecules. The receptors are integral membrane proteins, embedded in the lipid bilayer of the cell surface. T cells can be subdivided into **cytotoxic** T cells, which identify and kill foreign cells, **T-helper** cells, which stimulate B lymphocytes to produce high levels of antibody, and **delayed-typed hypersensitivity** T cells.

In addition to the **humoral immunity** conferred by the production of soluble antibodies, there is **cellular immunity** in which **activated** T lymphocytes are directly involved in eliminating foreign antigens. There are several mechanisms by which this is accomplished. A subset of T lymphocytes, the **cytotoxic T cells**, use specific surface receptors to recognize foreign antigens bound to MHC Class 1 antigens on cells; the T_c cells can lyse these cells when in contact with them. This mechanism is the primary one causing rejection of **tissue grafts**. **Natural killer cells** can kill foreign cells that they contact in the absence of stimulation by a specific antigen.

T cells release **lymphokines** when they bind to foreign antigens. These molecules modulate the activity of macrophages by attracting them to a site where foreign cells are present, stimulating their killing activity, and inhibiting their migration away from the antigen. The activation of macrophages occurs by stimulating the formation of hydrolytic granules during their development. These activated macrophages are more effective at intracellular killing of pathogens that normally multiply in the macrophage.

Thus, individual B lymphocytes are not identical, but are programmed to produce unique antibodies. Yet all these cells were derived from identical stem cells in the bone marrow. How is this diversity generated? It occurs by extensive rearrangements of genes during the maturation of the pool of B lymphocytes. The final transcriptional units coding for the light and heavy chains of a specific antibody in a mature B cell are assembled from several sets of genes. These include DNA coding for the following regions: constant, variable, joining, and (in the case of heavy chains) diversity. The DNA used for the constant region is the same for all B cells producing the same class of antibody, but the cells contain multiple forms of the sequences for the other regions. By genetic recombination, randomly selected variable, **joining**, and **diversity** segments are fused to produce the **active gene** in a particular B cell. The number of possible combinations is sufficient to generate 10 million unique B cells, each producing its own specific antibody. Additional diversity arises because the DNA splicing operations are imprecise, which create small shifts when joining the various segments into the active gene.

The antiserum collected from an animal challenged with an antigen contains a variety of antibodies which differ in class and specificity. This variety arose because a number of individual B cell clones were selected by the different antigenic determinants in the antigen. If a B cell clone were isolated from this mixture, the product would be antibodies of a single specificity, that is, a **monoclonal antibody**. This is accomplished by the **hybridoma** technique in which B cells isolated from an immunized animal are fused to **myeloma** cells so that they can be easily grown in tissue culture. Monoclonal antibodies react with only one antigenic determinant. Therefore, they can be used in research to separate complex mixtures of cells, or study the active site of enzymes. In medical diagnostics, they provide better resolution in the typing of bacteria or tissues. It may be possible to specifically target malignant cells in the body with them, and thereby deliver anticancer treatments specifically to tumor cells. Home pregnancy test kits are based on the detection of **human chorionic gonadotropin** (HCG) in urine using a monoclonal antibody system.

Because antibodies react very specifically with their corresponding antigens, they can be used in diagnostic labs to detect the presence of specific pathogens or types of tissue. Several types of **serological tests** have been developed for use in the laboratory. These include **neutralization** reactions, in which the binding of antibody blocks the biological or chemical activity of toxins, enzymes, or viruses. If the antigen is soluble, the bivalent antigen binding sites of antibodies can result in large aggregates of antigen and antibody which are observed as a **precipitin** reaction. If the antigens are particles, such as bacterial cells, antibodies directed against cell surface components cause large clumps to form, and this is known as the **agglutination** reaction.

Precipitation reactions can be carried out in agar gels. Antigen and antibody are placed in separate wells cut in the agar. Each diffuse out, and where the concentrations of the two substances are in **optimal proportions** a band of precipitate forms. Homologies between antigens can be discerned by observing what types of bands are formed when the antigens are in adjacent wells.

Blood typing is an example of the agglutination reaction. The major antigen group on human red blood cells is the **ABO system**. Individuals may have on their cells A antigen, B antigen, both or neither. The serum contains antibodies against the antigen(s) not present on the blood cells.

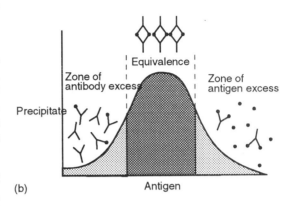

If more sensitivity is required to detect small amounts of antigen-antibody complexes, ELISA or radioimmunoassay can be used. In ELISA, an enzyme is covalently linked to an antibody, and the amount of antigen-antibody complex is quantified by adding the enzyme's substrate and measuring how much product is formed. In radioimmunoassay, radioactive iodine is linked to the antibody, and the amount of radioactivity that becomes associated with the antigen preparation is determined.

The immune system causes damage to host tissues in some instances. **Allergic** reactions are of two types: **delayed hypersensitivities**, due to T_{dth} cells and **immediate hypersensitivities**, which are caused by IgE antibody reactions.

Delayed hypersensitivity is evident a few hours to a few days after exposure to antigen. T_D cells bind to antigen and release lymphokines that attract macrophages and cytotoxic T cells to the site. These elicit a general inflammatory response which causes tissue damage. The **tuberculin skin test** is an example of a delayed hypersensitivity reaction, as are skin reactions to poison ivy and cosmetics.

Immediate hypersensitivities occur within a few hours of exposure to the antigen. The primary immune response to an allergen produced IgE antibodies, which attach to the surfaces of **mast cells** and **basophils**. Upon subsequent exposure to antigen, the binding of antigen to these antibodies triggers the release of **histamine** and **serotonin** from the cells. These molecules cause dilation of blood vessels and contraction of smooth muscle. The results are symptoms that can be mild (itching, watery eyes, or hives) to severe (respiratory distress and a drop in blood pressure). The severe symptoms can result in **anaphylactic shock**, which if untreated may cause death.

Under normal conditions, an immune response against the body's own molecules does not occur. As mature B and T cells are formed, the active genes for antibody or receptor are generated at random, so lymphocytes that could respond to host antigens are created. However, the exposure of an immature B cell to the self antigen during fetal development suppresses its development into an antibody-producing cell. T cell clones directed against self antigens are thought to be deleted by an unknown mechanism. In some cases, the **tolerance** mechanisms fail, and antibodies or immune cells against self antigens are made. These **autoimmune diseases** include juvenile diabetes, systemic lupus erythematosis, and multiple sclerosis.

The action of antibody upon foreign cells may involve the participation of a group of serum enzymes called **complement**. The complement system is activated by the binding of antibody to antigen and may (1) lyse cells, (2) kill without lysis, or (3) promote phagocytosis of encapsulated bacteria.

The complement system involves 11 proteins, which act in a sequential fashion. Only the reactions through C3 are necessary for **opsonization**, because phagocytes have C3 receptors on their surface. Further reactions lead to the attraction of leukocytes, and the action of the entire system results in damage to the bacterial cell membrane and lysis.

Immunological disorders in which the immune system does not respond to a foreign antigen include **agammaglobulinemia**, in which B cells are defective, and **AIDS**, in which a virus inhibits the activity of CD4 cells.

Specific immunity to protect against disease can be induced in two ways. (1) To produce **active immunity**, the individual is injected with antigen, to induce the formation of antibodies in the primary immune response. (2) Individuals exposed to the disease can be given **passive immunity** by injecting antibodies produced in another individual.

The **vaccine** used to induce active immunity must contain material that does not cause the disease, but does induce the formation of protective antibodies. Bacteria or viruses may be killed by heating or chemical treatment. However, live cells or virus are better immunogens, so attempts are made to isolate mutant **attenuated** strains that have lost their virulence for use in the vaccine. Exotoxins are chemically modified to nontoxic **toxoids** for incorporation into vaccines.

The **antiserum** administered for passive immunity is obtained from large, immunized animals or from **hyperimmune** humans.

SELF TESTS

COMPLETION

1. In clonal selection, the presentation of antigen by _____ results in selection of the B lymphocyte genetically programmed to make antibody that reacts with that antigen.

2. Low molecular weight substances that react with antibodies but do not induce antibody formation are _____.

3. The portion of a macromolecule to which a specific antibody binds is called the _____.

4. Immunoglobulins found in secretions are of the class _____.

5. The cells involved in the immune response which mature in the thymus are _____.

6. The major immune reaction against tissue grafted from one animal species to another is directed against _____ antigens.

7. The antigen receptors of T cells are part of the cell's _____.

8. The production of soluble antibody in the primary immune response requires T-_____ cells.

9. To prepare monoclonal antibodies, myeloma cells are fused with _____.

10. The serological reaction of antibodies with a particulate antigen is called _____.

11. An individual of blood type AB would possess natural antibodies against _____ blood type antigen(s).

12. The greatest sensitivity in serological reactions can be achieved using _____ methods.

13. If protective immunity can be transferred from one animal to another in whole blood, but is not transferred in serum, the protective factor is _____.

14. Immediate hypersensitivities are mediated by _____.

174

15. The absence of an immune response to one's own macromolecules is known as immune _____.

16. AIDS virus specifically inactivates _____ cells.

KEY WORDS AND PHRASES

active and passive immunity (p 862)	active immunity (p 862)
antibody (p 818)	antigen (p 818)
antitoxin (p 858)	attenuated strains (p 863)
bivalent antibody (p 830)	CD4 lymphocytes (p 827)
CD8 lymphocytes (p 827)	cell-mediated immunity (p 839)
clonal selection (p 844)	complement (p 850)
direct and indirect staining (p 856)	ELISA (p 856)
Fab fragment (p 830)	Fc fragment (p 830)
fluorescent antibody (p 856)	gamma globulin (p 829)
hapten (p 818)	humoral immunity (p 818)
hypersensitivity (p 860)	hypervariable regions (p 846)
immunogen (p 818)	immunoglobulin (p 829)
immunoglobulin gene superfamily (p 849)	latent period (p 843)
line of identity (p 857)	line of partial identity (p 857)
monoclonal antibodies (p 852)	passive immunity (p 862)
plasma cells (p 844)	polyclonal antibodies (p 852)
spur (p 857)	

MULTIPLE CHOICE

1. Infectious diseases caused by organisms that produce exotoxins are prevented by immunization with toxoids because:

    (A) toxoids are the virulence factors of these organisms

    (B) the bacteria grow in areas of the body which antibodies cannot reach

    (C) the toxin is responsible for disrupting body function rather than the bacterial cells themselves

    (D) toxoid can be produced in large quantities and bacterial cells cannot

    (E) toxoids are better antigens than the bacterial cells

175

2. Foreign substances that induce a specific immune response after gaining entrance to the body are:

    (A) antibodies   (D) immunogens

    (B) complement   (E) lymphokines

    (C) haptens

3. All of the following statements about an IgG molecule are TRUE EXCEPT:

    (A) There are four polypeptide chains joined by disulfide bonds.

    (B) There are two antigen-binding sites.

    (C) The forces involved in antigen-antibody binding are weak bonds rather than covalent bonds.

    (D) If all disulfide bonds joining the polypeptides are broken, the result is two identical light chains and two identical heavy chains.

    (E) The constant regions of the heavy and light chain have identical sequences.

4. Which of the following is not true of immediate hypersensitivity?

    (A) Involves IgE antibodies

    (B) Occurs within a few hours after exposure to antigen

    (C) Involves the inflammatory response

    (D) Involves cell-mediated immunity

    (E) Mild attacks are treated with anti-histamines

5. Antigen and antibody molecules will form a visible precipitate if they:

    (A) interact in a cross-linked lattice formation

    (B) stimulate lymphokine production

    (C) attract phagocytes

    (D) neutralize viruses

    (E) include IgE antibodies

6. A signal involved in triggering the proliferation of a specific B lymphocyte (that is, the primary immune response) is:

    (A) the release of lymphokines from a cytotoxic T cell

    (B) the interaction of antigen on the surface of a macrophage with a receptor on the B lymphocyte

    (C) a hormone produced in the bursa

    (D) histamine

    (E) the lysis of macrophages

7. The titer of specific antibody rises within a few days after the second exposure to an antigen because

   (A) the body senses that the introduction of the antigen was not an immunization, but rather a virulent pathogen from nature

   (B) a previous experience with the antigen produced memory cells which can respond quickly to subsequent challenge with antigen

   (C) the body's other defense mechanisms are better equipped to fight the foreign substance because of the previous exposure

   (D) high concentrations of antibody were synthesized after the first exposure and stored in the thymus. These reserves are released when the antigen is encountered again

8. Circulating immunoglobulins are formed and excreted by:

   (A) macrophages   (D) memory cells

   (B) basophils   (E) leukocytes

   (C) plasma cells

9. Tumor cells that arise in the body will usually be recognized and destroyed by T lymphocytes before a large cancer develops because:

   (A) tumor cells have unique antigens which are perceived as foreign

   (B) these cells elicit an immediate hypersensitivity reaction

   (C) the tumor cells secrete lymphokines

   (D) IgE antibodies bind to tumor cells and mark them as cancerous

   (E) the tumor cells produce toxins which attract cytotoxic T lymphocytes

10. The normal adult human being can respond immunologically to:

   (A) protein antigens only

   (B) polysaccharide antigens only

   (C) most substances recognized as "self"

   (D) non-human substances only

   (E) millions of different antigenic determinants

11. The interactions of antigens and antibodies have all of the following characteristics EXCEPT:

   (A) Depend on the formation of weak noncovalent chemical bonds

   (B) A number of bonds must be formed

   (C) Require a close conformational fit

   (D) It is a reversible reaction

   (E) The presence of a non-protein co-antibody is required

12. In immunization, an inactivated pathogenic microbe is injected as an antigen. This provides long-lasting immunity because:

   (A) plasma cells continue to produce large quantities of specific antibody for a number of years

   (B) a high concentration of lymphokines are maintained in the serum

   (C) memory cells are produced which can respond very quickly to subsequent introductions of antigen

   (D) macrophages are activated

   (E) the antigen causes gene rearrangements in the variable regions of the immunoglobulin genes

13. All of the following are involved in cell-mediated immunity EXCEPT:

   (A) B lymphocytes   (D) lymphokines

   (B) T lymphocytes   (E) natural killer cells

   (C) macrophages

14. In preparing a vaccine, the toxicity of exotoxin can be reduced by:

   (A) treating the toxin with formaldehyde

   (B) administering an antibiotic with the toxin

   (C) filtration to eliminate the toxic component

   (D) any of the above

   (E) none of the above

15. Human chorionic gonadotropin (HCG) is produced in response to what process:

   (A) B-cell interaction with an antigen

   (B) ovum fertilization

   (C) a non-functioning class I MHC proteins

   (D) any of the above

   (E) none of the above

MATCHING (Match the autoimmune disease with the affected system)

Disease

   1. juvenile diabetes
   2. myasthernia gravis
   3. rheumatoid arthritis
   4. pernicious anemia
   5. Addison's disease
   6. multiple sclerosis
   7. Goodpasture's syndrome
   8. Hashimoto's disease
   9. lupus erythematosis

System

   (A) brain
   (B) adrenal glands
   (C) DNA, others
   (D) intrinsic factor
   (E) kidney
   (F) pancreas
   (G) skeletal muscle
   (H) cartilage
   (I) thyroid

DISCUSSION

1. Distinguish between humoral and cellular immunity in terms of (a) the entity responsible for immunity, (b) the specificity of immunity, and (c) the fate of the antigen.

2. What types of molecules can function as immunogens? What determines the number of different antibodies made against an immunogen?

3. What characteristics distinguish the five classes of immunoglobulin molecules? What properties do they have in common?

4. Draw the structure of an IgG molecule and indicate the following features: heavy chain, light chain, constant regions, variable regions, antigen-combining sites, binding site for C1 complement protein.

5. What immunological functions are performed by the following sets of T lymphocytes: T_h, T_{dth}, and T_c?

6. What role do macrophages play in initiating a humoral immune response?

7. How do the major histocompatibility complex antigens interact with T lymphocytes to facilitate immune responses?

8. Why does the secondary response to an antigen occur more rapidly and produce a higher titer of antibody than the primary antibody response?

9. What sets of genes are involved in generating the genes for light and heavy chains in a mature B lymphocyte? By what mechanisms can a pool of B lymphocytes producing millions of different antibodies be generated from genetically identical stem cells?

10. How do monoclonal and polyclonal antibodies differ in their reaction with an antigen, such as a bacterial cell? What steps in preparing the antibody are similar for each type? What steps are unique to the preparation of monoclonal antibodies?

11. Describe three types of reactions by which the formation of antigen-antibody complexes can be detected. Which method(s) are most sensitive? What methods are most rapid and easy to carry out?

12. By what mechanisms can T lymphocytes directly or indirectly eliminate foreign cells from the body?

13. When tissue is transplanted from one person to another, the tissue cells are often killed. By what mechanism are cells killed, what components of the tissue is recognized as foreign, and how might one minimize the chances of rejection of foreign tissue?

14. Distinguish between immediate and delayed hypersensitivity in terms of (a) the immune system that is involved, (b) the time required, and (c) the type of reactions responsible for tissue damage.

15. Why don't humans generally mount either a humoral or cell-mediated immune response against the macromolecules that constitute the body?

16. How is the complement system activated? What effects can complement activation have upon foreign cells?

17. What material would you use to immunize (a) an individual who had ingested botulinum toxin and (b) to protect an individual who might be inoculated with *Clostridium tetani* following a puncture wound?

ANSWERS

Completion

1. macrophages; 2. haptens; 3. antigenic determinant; 4. IgA; 5. T lymphocyte; 6. major histocompatibility complex; 7. cytoplasmic membrane; 8. helper; 9. B lymphocyte; 10. agglutination; 11. neither; 12. ELISA or radioimmunoassay; 13. T lymphocytes; 14. IgE antibodies; 15. tolerance; 16. CD4.

Multiple choice

1. C; 2. D; 3. E; 4. D; 5. A; 6. B; 7. B; 8. C; 9. A; 10. E; 11. E; 12. C; 13. A; 14. A; 15. B.

Matching

1. F; 2. G; 3. H; 4. D; 5. B; 6. A; 7. E; 8. I; 9. C.

Discussion

1. See text Sections 20.1, 20.3, 20.6, 20.10 and 20.12
2. See text Section 20.1
3. See text Section 20.3
4. See text Section 20.3
5. See text Sections 20.2, 20.6, 20.8, and 20.14
6. See text Section 20.8
7. See text Section 20.4
8. See text Section 20.8
9. See text Section 20.9
10. See text Section 20.11
11. See text Sections 20.12 and 20.13
20. See text Section 20.6
13. See text Sections 20.5 and 20.13
14. See text Sections 20.6 and 20.14
15. See text Section 20.7
16. See text Section 20.10
17. See text Sections 20.14 and 20.15

Chapter 21
CLINICAL AND DIAGNOSTIC MICROBIOLOGY AND IMMUNOLOGY

OVERVIEW

Chapter 21 (pages 869-905) describes the principle of clinical and diagnostic microbiology as used to isolate and identify disease causing agents. The diagnosis and treatment of infectious diseases can be expedited by the rapid identification of the responsible microbe. The traditional techniques of isolation, identification, and antibiotic susceptibility testing take up to 72 hours. However, more rapid procedures are being developed to decrease this time, and some of these approaches are discussed in the chapter. In addition, the factors which are important in sample collection, cultivation, and identification of pathogens are presented.

CHAPTER NOTES

Samples that are collected from infected tissues and fluids must be taken as aseptically as possible, and should be analyzed promptly to prevent chemical or biological changes during storage. There is always a chance that samples from the body have been contaminated with the normal flora during collection. These instances can be identified by considering what species have been isolated. If it is a member of the normal flora which resides in the tissues near sample collection, it probably was a contaminant. Special protocols are used to sample blood, urine, fecal material, wounds and abscesses and the genital area. In the case of **urinary tract infections**, the number of bacteria in the sample gives an indication to whether or not the sample is contaminated; direct microscopic counts are of value here.

Perhaps the most important tool in the isolation of pathogens from samples taken from infected individuals is the use of **restrictive growth conditions**. For example, incubating samples under anaerobic conditions will obviously prevent the growth of contaminating obligate aerobes. If some unique characteristics of the suspected pathogen are known, these can be incorporated into the medium design. For example, **Thayer-Martin agar** contains several antibiotics to which *Neisseria gonorrhoeae* is resistant. This is an example of a **selective medium**, which contains compounds that inhibit unwanted microbes but not the bacteria of interest.

Differential media are useful for the identification of isolated bacteria, because they permit reactions such as acid production, red blood cell hemolysis, or hydrogen sulfide production to be conveniently visualized. In actual practice, many media have both selective and differential properties.

To complete the identification of an isolate, a bank of differential media and biochemical tests can be used (see Table 13.2 and 13.3). These tests assay for the presence of specific enzymes under the growth conditions that are provided. By comparing the results obtained from an unknown isolate to data for known species, an identification can be made. The most useful tests for prevalent pathogens are packaged in commercial kits, so that small laboratories need not prepare a variety of specialized media.

The sensitivity of an isolated pathogen to clinically useful antibiotics is assayed by the **Kirby-Bauer method** (see figure 13.6). A culture of the bacterium is swabbed across an agar plate, and paper discs containing different antibiotics are placed on the plate. The antibiotics diffuse out into the agar. If the bacterium is resistant to an antibiotic, a lawn of bacteria will grow right up to the disc. However, if an antibiotic inhibits growth, a clear zone without bacterial growth will result around the disc; the size of the zone is proportional to the sensitivity of the bacterium to the drug.

The **minimum inhibitory concentration** of an antibiotic for a bacterial species is determined by inoculating a series of tubes containing different antibiotic concentrations, and observing the lowest concentration at which no growth occurs.

If an **immunological method** can be used directly upon a clinical specimen, a pathogen can be identified without culturing it. This requires the preparation of **reference antisera** against known pathogens (also see text 12.11). **Fluorescent antibodies** can be visualized microscopically after binding to a specimen. However, the **specificity** of polyclonal antibodies may be a problem. Members of the normal bacterial flora may have similar surface antigens to the pathogen and **cross-react** with the reference antiserum. This problem can be overcome by the use of **monoclonal antibodies**. If a monoclonal antibody reacts with a very unique surface component, it may only detect one particular strain of bacterium. In contrast, some antigenic determinants are similar in different species. A monoclonal antibody that reacted with this determinant could be used in a broad screening program.

ELISA assays have become popular in diagnostic labs because they can be automated and used in mass screening programs (also see text 12.13). **Direct ELISA** assays are used to detect antigens in a patient specimen, whereas **indirect ELISAs** can detect specific antibodies in patient serum. For example, the test for exposure to HIV virus (the cause of AIDS) is an indirect ELISA that detects anti-HIV antibodies. The most important advantage of ELISA methods is their **sensitivity**, their ability to detect small amounts of antigen-antibody complexes. This sensitivity allows an early detection following exposure to a pathogen

Agglutination tests can be done quickly and inexpensively. By coating antigen or antibody on latex beads, a positive reaction can be easily seen. The technique is inexpensive; thus it is especially useful for large scale screening programs or in developing countries where specialized equipment is difficult to maintain.

Clinical specimens contain a complex mixture of antigens. To detect specific **protein** antigens, a **Western blot** can be performed. In this technique, proteins are separated on the basis of size by **polyacrylamide gel electrophoresis**. The proteins are transferred to nitrocellulose paper. Antibody solution is added to the paper, and then molecules which detect antigen-antibody complexes are added. If **protein A** is used, it contains radioactive iodine which exposes X-ray film placed over the blot at spots where antigen-antibody complexes had formed. Alternatively, an anti-immunoglobulin antibody linked to an enzyme can be used. When enzyme substrates are added, a colored precipitate is deposited wherever the enzyme was fixed to the blot.

Almost all bacteria contain **plasmids**, but the type and number can differ between species. These differences can be analyzed by **DNA fingerprinting** in which a cell lysate is subjected to agarose gel electrophoresis to separate plasmids. The pattern obtained is diagnostically useful in determining whether clinical isolates are related to one another.

A more specific technique is the use of **nucleic acid probes**. If a DNA sequence unique to a pathogen can be isolated, its **hybridization** to DNA from an unknown isolate can be used to identify the isolate. An example is a probe which contains part of the *E. coli* enterotoxin gene sequence. This can be used to distinguish pathogenic from nonpathogenic strains.

The probe technique is sensitive enough to detect the DNA contained in 1000 cells. Thus, time can be saved by using it directly in clinical specimens so that the bacterium need not be grown in culture. For detection purposes, the probe must contain a **reporter molecule**. The reporter can be a radioisotope, a fluorescent molecule, or an enzyme.

Viral pathogens are difficult to culture routinely in the laboratory. Therefore, for common infections, the diagnosis depends upon the physician's observation of symptoms or upon serological laboratory reactions.

Medical microbiologists must take special care to prevent infecting themselves or others with pathogens contained in samples.. A major mechanism of infection is through **aerosols**, which can be formed during pipetting or centrifugation. For extremely hazardous pathogens, the filtration and air flow in the laboratory may be specially designed to prevent the escape of microbes from the laboratory.

SELF TESTS

COMPLETION

1. For accurate analysis, clinical specimens must be collected under _____ conditions.

2. An infection in the blood is called a _____.

3. The bacterial count in infected urine samples can exceed _____ cells per ml.

4. Gonorrhea is diagnosed by examining clinical specimens consisting of _____.

5. Clinical samples which will be tested for obligate anaerobes must be collected in a tube containing _____.

6. Indicator dyes sensitive to pH changes are added to fermentation broth to detect the production of _____.

7. Methylene blue inhibits the growth of _____ bacteria.

8. The oxidase reaction tests for the presence of a specific _____.

9. In the agar diffusion method, the sensitivity of a bacterium to an antibiotic is recognized by a _____ in the agar.

10. To avoid carrying pathogens out of a clinical laboratory, one should wear _____ while working in the lab.

11. _____ can be used to microscopically detect specific pathogens in a clinical specimen.

12. Clinical immunological identification of *Vibrio cholerae* is based on the _____ antigen.

13. In the latex bead agglutination procedure, _____ or _____ are coated on the outside of the bead.

14. In the diagnosis of nonbacterial meningitis, it is necessary to obtain _____ from the patient.

15. Precipitation reactions between antibodies and and a soluble antigen will result in the formation of _____ complex.

KEY WORDS AND PHRASES

agglutination (p 885)	AIDS (p 900)
anaerobes (p 875)	bacteremia (p 870)
carbohydrate fermentation (p 877)	chocolate agar (p 874)
clinical/diagnostic microbiology (p 870)	coagulase test (p 878)
detecting antibodies (p 883)	differential medium (p 876)
ELISA (p 884	enteric bacteria (p 878)
eosin-methylene-blue EMB (p 876)	fecal culture (p 873)
fluorescent antibody (p 883)	gonorrhea (p 874)

growth-dependent diagnostics (p 876)	HeLa cells (p 901)
HIV virus (p 892 & 900)	Kirby-Bauer method (p 880)
monoclonal antibody (p 884)	nosocomial (p 871)
nosocomial infection (p 872)	nucleic acid probe (p 898)
oxidase test (p 878)	plasmid fingerprinting (p 898)
polyclonal antibody (p 884)	selective medium (p 876)
septicemia (p 870)	titer (p 883)
triple sugar iron agar (p 879)	urinary tract infection (p 872)
Voges-proskauer test (p 879)	

MULTIPLE CHOICE

1. Urinary tract infections are most commonly caused by:

    (A) staphylococci   (D) clostridia

    (B) streptococci   (E) ureolytic bacteria

    (C) enteric bacteria

2. A selective medium for the growth of gonococci is:

    (A) blood agar   (D) Thayer-Martin agar

    (B) chocolate agar   (E) nutrient agar

    (C) EMB agar

3. EMB agar is a medium which is:

    (A) differential

    (B) non-selective

    (C) selective

    (D) selective and differential

4. Ferric iron is put into media to detect the production of:

    (A) transferrin

    (B) antibiotics

    (C) organic acids

    (D) catalase

    (E) hydrogen sulfide

5. Carbohydrate fermentation tests are used to differentiate:

    (A) Gram positive bacteria

    (B) enteric bacteria

    (C) obligate anaerobes

    (D) gonococci

    (E) clostridia

6. Antibiotic sensitivity of a clinical isolate is tested by the:

    (A) Voges-Proskauer test

    (B) Thayer-Martin method

    (C) ONPG test

    (D) Kirby-Bauer method

    (E) streptomycete assay

7. Human viruses are typically grown in diagnostic virology labs using all of the following types of cells EXCEPT:

    (A) HeLa cells

    (B) WI-38 fibroblasts

    (C) hybridomas

    (D) Rhesus monkey kidney cells

8. Which of the following procedures should not be allowed in a clinical laboratory?

    (A) drinking water

    (B) inviting friends to visit the lab

    (C) pipetting by mouth

    (D) disposing of used samples in the public trash

    (E) all of the above

9. A potential disadvantage of immunological techniques in clinical laboratories is that they may be:

    (A) nonspecific   (C) time-consuming

    (B) insensitive   (D) expensive

10. Indirect ELISA assays are used in clinical laboratories to detect:

    (A) pathogens in clinical specimens

    (B) antibodies in patients' serum

    (C) cell surface components of transplanted tissue

    (D) protein A

    (E) nucleic acid probes

11. A specific gene characteristic of a microbial pathogen can be identified in a clinical specimen by using a:

    (A) fluorescent antibody

    (B) nucleic acid probe

    (C) Western blot

    (D) direct ELISA assay

    (E) catalase test

12. Use of bismuth sulfate agar (BS) to isolate *Enterobacter* and *Salmonella* bacteria is of particular value as:

   (A) *Enterobacter* has mucoid colonies with silver sheen and *Salmonella* has colonies that are black to dark green when grown on BS

   (B) *Enterobacter* is mostly inhibited and *Salmonella* has colonies that are black to dark green when grown on BS

   (C) *Enterobacter* has mucoid colonies with silver sheen and *Salmonella* is mostly inhibited when grown on BS

   (D) *Enterobacter* has brown colonies with a gold sheen and *Salmonella* has colonies that are gray when grown on BS

   (E) BS agar is not useful

13. Nucleic acid hybridization is a powerful tool for the detection of specific organisms. Why is it such a specific indicator?

   (A) the presence of specific DNA sequences belonging to specific bacteria can be identified

   (B) the probe technology is very sensitive, allowing detection of low levels of bacteria

   (C) nucleic acids are more stable than proteins allowing a harsh handling regime

   (D) none of the above

   (E) all of the above

14. Plasmid fingerprinting is considered as an adjunct or complementary method because:

   1. the presence or absence of plasmids may have little to do with taxonomic definition of the organism

   2. plasmids may transfer into or out of a test strain, changing it as compared to reference strain

   3. it is overly sensitive and often detects all DNA

   4. it is indirect and requires many steps

   5. identical plasmid profiles are only suggestive of relatedness of the organisms

   (A) 1, 3, 4; (B) 1, 2, 5; (C) 3, 4, 5; (D) 1 5; (E) 3 4.

15. Eosin-methylene blue (EMB) is one of the most widely used selective and differential medium. EMB will allow the selection of Gram-negative enteric bacteria because:

   1. methylene blue inhibits all Gram-positive and Gram-negative bacteria except enteric bacteria

   2. eosin changes colors under acid conditions, thus acting as an indicator of lactose fermentation

   3. only enteric bacteria can use lactose and sucrose as carbon sources

   4. methylene blue inhibits Gram-positive bacteria

   (A) 1, 3, 4; (B) 1, 2, 3; (C) 2, 4, ; (D) 1, 2; (E) 3, 4, .

MATCHING (Match the clinical test with it operating principle)

TEST   **PRINCIPLE**

1. carbohydrate fermentation   (A) acetoin from sugar fermentation

2. catalase

(B) urea spilt to NH_3 and CO_2

3. citrate utilization

(C) cytochrome c oxidizes electron acceptor

4. coagulase

(D) enzymatic breakdown of H_2O_2

5. indole test

(E) CO_2 production from amino acid

6. methyl red test

(F) mixed-acid fermentation lowers pH to growth on sugars

7. decarboxylases of amino acids

(G) tryptophan converted to indole

8. gelatin liquefaction

(H) use of citrate as sole carbon source

9. nitrate reduction

(I) clotting of blood

10. oxidase test

(J) proteases breakdown gelatin

11. urease test

(K) NO_3^- reduced to NO_2^-

12. Voges-proskauer test

(L) acid production from aerobic growth

13. oxidation-fermentation

(M) acid and or gas from fermentative growth on sugars

DISCUSSION

1. Why is it important to analyze a clinical specimen as soon as possible after it is collected?

2. What precautions should be taken in collecting a clinical specimen? What additional precautions are necessary if the suspected pathogen is an obligate anaerobe?

3. What techniques are used to diagnose gonorrhea?

4. Differentiate between selective and differential media. For what purpose is each type used?

5. List three types of metabolic products and three specific enzyme reactions which are assayed in diagnostic tests.

6. Why is the Kirby-Bauer method a more efficient method of determining the antibiotic susceptibility pattern of a clinical isolate than the antibiotic dilution assay?

7. What procedures should be followed in a clinical diagnostic laboratory to minimize the risk of laboratory-acquired infections?

8. Describe two techniques which can be used to identify pathogens directly in clinical specimens, without first growing them in culture.

9. What diagnostic tests are used to detect the presence of antibodies to HIV virus in human serum?

10. What advantages do ELISA assays have over conventional serological tests in the clinical diagnostic lab?

11. In what types of immunological and nucleic acid diagnostic tests are enzymes used to quantify the reaction?

ANSWERS

Completion

1. aseptic; 2. septicemia; 3. 105; 4. urethral discharge or vaginal smear; 5. oxygen-free gas; 6. acid; 7. Gram positive; 8. cytochrome; 9. zone of inhibition; 10. a lab coat; 11. fluorescent antibodies; 12. O; 13. antibodies or antigens; 14. spinal fluid; 15. visible and insoluble.

Multiple choice

1. C; 2. D; 3. D; 4. E; 5. B; 6. D; 7. C; 8. E; 9. A; 10. B; 11. B; 12. A; 13. E; 14. B; 15. C.

Matching

1. M; 2. D; 3. H; 4. I; 5. G; 6. F; 7. E; 8. J; 9. K; 10. C; 11. B; 12. A; 13. L.

Discussion

1. See text Section 21.1

2. See text Section 21.1

3. See text Section 21.1

4. See text Section 21.2

5. See Table 21.3

6. See text Section 21.3

7. See text Section 21.12

8. See text Sections 21.5 and 21.10

9. See text Sections 21.7 and 21.9

10. See text Section 21.7

11. See text Sections 21.7, 21.9, and 21.10

EPIDERMIOLOGY AND PUBLICH HEALTH MICROBIOLOGY

OVERVIEW

Chapter 22 (pages 906-932) concerns itself with microbial detective work:, the investigation of how pathogens spread through populations, and how this spread can be controlled. An **epidemiologist** can use statistical analysis of data on diseases to track the spread of pathogens. The insights obtained through the science of microbiology have eliminated microbial diseases as major causes of death in developed countries. However, the inability to effectively apply public health controls and a poorer standard of living makes infectious diseases a major cause of mortality in developing countries.

CHAPTER NOTES

In Chapter 19, the interaction between a pathogen and its host was described as dynamic, because the pathogen's virulence factors attempted to circumvent host defense mechanisms. Both pathogen and host are capable of genetic change, and hence can **coevolve** in response to one another. For this reason, well-adapted parasites generally do not kill their hosts, because the host is a good environment to maintain the parasite. Lethal pathogens arise due to a genetic change in the pathogen to which the host cannot respond, or when the host is extremely susceptible to the pathogen.

Describing how a disease is distributed within a population is the work of the epidemiologist. The **incidence** describes the number of diseased individuals within the population as a whole. A diseases is described as being at **epidemic** levels if it occurs in a high number of the population. A **pandemic** occurs when the epidemic becomes widespread. A disease is said to be **endemic** if it occurs at low levels in a population. A typical disease infection process follow a common set of stages: the initial growth in the host is termed **infection**, the time between the infection and when the first symptoms occur is termed the **incubation period**, the initial infection can also be followed by a **prodromal** phase when first symptoms occur. When the disease is most active it is referred to as being **acute**. Following the acute period is the period of **decline** when the fever subsides and the infected individual begins to return to normal. The time between the decline period and the complete regaining of health is termed the **convalescent** phase.

Two important determinants of how a disease will spread are the **reservoir** of the pathogen, and its **mode of transmission**. The **reservoir** for some pathogens is soil or water, where they are normal inhabitants; they only incidentally cause human disease. Little can be done to eradicate such pathogens. However, many pathogens can only exist in living organisms. If humans are the reservoir, person-to-person contact is required for disease transmission. Carriers are especially important reservoirs of human pathogens, because the individual may not recognize that he is disseminating pathogens. Identification of carriers who can infect large groups of individuals is an important method of preventing the spread of some diseases. **Zoonoses** are diseases which primarily occur in animals, but are occasionally transmitted to humans. Elimination of the disease in the animal reservoir by immunizing susceptible animals and slaughtering infected ones is the key to preventing its occurrence in humans. Several pathogenic protozoa have complex life cycles which involve different stages in a human and in another animal.

The investigation of AIDS is a good case study of how epidemiological research can be applied. Statistical analysis indicated that the disease was associated with certain **high-risk**

groups. The association with specific groups suggested that transmission did not occur by casual contact or through air or water, because then the population at large would be more widely infected.

The mode of **transmission** of a pathogen is usually related to its habitats in the body. Respiratory pathogens are transmitted by aerosols in the air, whereas intestinal pathogens are ingested in contaminated food or water. Transmission requires mechanisms for (1) escape from the host, (2) travel to and (3) entry into a susceptible individual.

Direct transmission occurs with respiratory pathogens through infectious droplets released by coughing or sneezing. The pathogens responsible for **sexually-transmitted diseases** are so sensitive to drying that they do not survive outside the body; therefore, they must be transmitted by intimate contact.

Indirect transmission between hosts can be mediated by living agents, called **vectors**. Arthropods and vertebrates are important vectors for human diseases; the pathogen may be transmitted by biting. The vector may not be affected by carrying the pathogen. Food and water are important inanimate **vehicles** for the spread of disease epidemics. A small group of individuals can become infected by contact with fomites such as eating utensils.

Vehicles such as contaminated water can cause **common-source epidemics**, in which a large number of individuals are infected within a restricted period of time. These outbreaks rise sharply and decline rapidly. Outbreaks of cholera are an example of a common-source epidemic. In contrast, if only a few individuals are infected, and an incubation period must elapse before they infect other susceptible persons, a **propagated epidemic** occurs in which the disease incidence rises slowly and declines gradually.

The genetic changes which occur in host and pathogen populations as a consequence of **selection pressures** are illustrated by the rabbit-**myxoma virus** example. The rabbit population was initially highly susceptible to viral infection, but those rabbits who through genetic variation developed some resistance became a major proportion of the population, because most of the others were killed by the virus. During this time, the virus also became less virulent.

Immunization is an effective means of protecting individuals against a pathogen. But immunization of most of a population also protects those individuals who have not been immunized. This **herd immunity** occurs because the probability that a susceptible individual will encounter a pathogen is reduced by the low density of pathogens in the immunized population. The critical proportion of immune individuals necessary to protect the whole population depends upon the virulence and infectivity of a particular pathogen.

Disease control measures are especially important in hospitals, because many conditions which promote disease spread are present. Virulent pathogens are constantly introduced into the hospital by incoming patients. Hospital personnel can function as **carriers** to spread pathogens throughout the hospital. Some hospital procedures risk introducing pathogens into the patient. The patients are especially susceptible to infection because of disease, the stress of medical procedures, or drug treatment. Important **nosocomial** pathogens are *Staphylococcus aureus*, because it can be part of the normal flora in carriers and requires good sanitation procedures to remove it from inanimate objects, and *Pseudomonas aeruginosa*, because it is very resistant to antibiotic treatment.

The decreased mortality due to infectious diseases in the U.S. is a consequence of a higher standard of living, better sanitation procedures, and immunization. However, sexually transmitted diseases have increased in prevalence, due to changes in sexual behavior.

Some diseases can be controlled by attacking the reservoir. This has worked best when domestic animals are the reservoir, because the population can be immunized. This is more difficult with wild animal reservoirs. Controlling diseases with human reservoirs has been difficult, especially if there are asymptomatic carriers who are difficult to identify.

190

Water purification and quality control of food have decreased the transmission of many intestinal diseases. It is difficult to implement similar measures to prevent transmission of respiratory pathogens.

Widespread immunization has been responsible for controlling the incidence of several diseases. The proportion of the U.S. population immunized against some diseases is falling, because the diseases are incorrectly viewed as no longer being risks. In addition, adults who were immunized as children may have inadequate protection, because the level of immunity falls if the system is not periodically challenged with antigen.

Infectious diseases still account for 30-50% of deaths in developing countries. Most of these deaths occur in children. Nutritional deficiencies and poorer sanitation procedures are responsible for the relatively high levels of infectious disease mortality in these countries.

SELF TESTS

COMPLETION

1. Mortality from infectious disease is _____ prevalent today in the U.S. than in 1900.

2. A worldwide epidemic is called a _____.

3. Individuals who are infected with a pathogen but do not show disease symptoms are called

   _____.

4. The major causes of illness can be determined by examining statistics on

   _____.

5. The reservoir of diseases known as zoonoses is _____.

6. Bovine tuberculosis has been controlled by _____.

7. Individuals who are serologically positive for HIV virus, but do not have clinical symptoms of AIDS are classified as having the condition_____.

8. AIDS is an example of a _____ epidemic.

9. The few non-vaccinated individuals in a population that is 95% immunized are protected by

   _____.

10. Hospital personnel may spread infectious disease because they are _____

11. *Pseudomonas aeruginosa* infections are difficult to treat because the organisms are

   _____

12. _____ is the study of the occurrence and spread of disease.

13. In the year _____, John Snow showed that cholera was transmitted by _____.

14. The occurrence of an infectious disease in hospital patients is termed _____ infection.

15. When inanimate objects transmit a disease causing organism, the object is termed a

   _____.

KEY WORDS AND PHRASES

epidemiology (p 908)	endemic disease (p 909)
epidemic (p 909)	pandemic (p 909)
disease reservoir (p 909)	subclinical infection (p 909)
carrier (p 909)	cycles of disease (p 918)

mortality (p 909)	morbidity (p 909)
zoonosis (p 910)	AIDS-related complex (p 913)
HIV virus(p 913)	vaccination (p 922)
nosocomial infection (p 920)	public health (p 921)
quarantine (p 923)	herd immunity (p 919)
propagated epidemic (p 917)	decline peroid (p 910)

MULTIPLE CHOICE

1. A species that is obligately parasitic will increase its chances of survival when it:

    (A) kills its host quickly

    (B) becomes less virulent, so that the host barely suffers from its presence

    (C) cannot be easily transmitted from host to host

2. Reservoirs of infectious diseases of humans include all of the following EXCEPT:

    (A) man   (D) water

    (B) animals   (E) All of these are reservoirs

    (C) soil

3. Outbreaks of infectious diseases are more likely to occur in

    (A) poverty-stricken areas

    (B) war-torn areas

    (C) developed countries

    (D) A and B above

    (E) B and C above

4. Epidemics of typhoid fever have been eliminated in the U.S. because of:

    (A) control of mosquito populations

    (B) treatment of sewage and water supplies

    (C) a decrease in virulence of Salmonella typhi

    (D) antibiotic therapy

5. Which of the following characteristics of the hospital environment does NOT increase the risk of nosocomial infections?

    (A) Virulent pathogens are constantly introduced.

    (B) The hospital staff may be carriers.

    (C) Certain therapies hamper the host's defense mechanisms.

    (D) Hospital floors are routinely swabbed with a wet mop.

    (E) There is a high population density.

6. All of the following are effective methods of preventing disease epidemics EXCEPT:

   (A) sewage treatment

   (B) immunization

   (C) identification of human carriers

   (D) improving the economic condition of the poor

   (E) all of the above are effective

7. Five babies have died of *Staphylococcus aureus* infections in one hospital nursery within six months. How could you find out whether these were the same or different strains?

   (A) DNA fingerprinting   (D) Gram stain

   (B) colonial morphology   (E) Any one of the above would be adequate

   (C) biochemical tests

8. Respiratory tract infections are most often transmitted by:

   (A) contaminated food   (D) direct oral contact

   (B) contaminated air   (E) arthropods

   (C) contaminated water

9. All of the following are at risk for AIDS EXCEPT:

   (A) hemophiliacs

   (B) sexually active bisexual men (with multiple sex partners)

   (C) sexually active homosexual men (with multiple sex partners)

   (D) sexually active heterosexual men and women (with multiple sex partners)

   (E) all of the above are at some risk

10. Which of the following diseases is only transmitted by direct contact between humans?

   (A) gonorrhea   (C) rabies

   (B) typhoid fever   (D) diphtheria

11. Which of the following could be a disease vector?

   (A) water   (D) eating utensils

   (B) soil   (E) surgical needles

   (C) mosquitoes

12. An ideal parasite will do all of the following EXCEPT:

   (A) grow in the host

   (B) live in harmony with the host

   (C) spread to other members of the population

   (D) reproduce in the host

   (E) kill the host

13. Mortality and morbidity refer to what two concerns:

   (A) disease and death in a population, respectively

   (B) death and disease in a population, respectively

   (C) the rate at which deadly diseases are affecting a community

   (D) the rate of death from fatal diseases

   (E) none of the above

14. The control of dysentery (*Shigella dysenteriae*) and cholera (*Vibrio cholerae*) within a community have a common approach:

   (A) decontamination of public water supplies

   (B) vaccination of individuals in the area

   (C) vaccination of livestock

   (D) vaccination of food handlers only

   (E) pasteurization of milk

15. What concept is used to describe the finding that 100% immunization is not necessary in order to prevent a disease from spreading in a population?

   (A) endemic immunity

   (B) public health

   (C) herd immunity

   (D) cycle of disease

   (E) endemic health

DISCUSSION

1. Is there any advantage to a pathogen quickly killing its host? Are there any disadvantages to doing so?

2. What procedures would one use to identify carriers of a specific disease?

3. Which diseases would be most difficult to eradicate -- those whose reservoirs are humans, wild animals, domestic animals, or soil?

4. Why was the incidence of AIDS much higher in cities like New York and San Francisco than in the rest of the country?

5. Name two types of infectious diseases that are transmitted directly from host-to-host.

6. How do insects transmit disease from an infected individual to a susceptible one?

7. Gonorrhea occurs in epidemic proportions in the U.S. Is this a common-source or a propagated epidemic? Does this mean that the number of gonorrhea cases will fall sharply or gradually if the disease were brought under control?

8. What changes in the host and pathogen populations occurred after the introduction of myxoma virus into Australia?

9. What factors increase the risk of infection in a hospital?

10. What public health measures have been used to control (a) the reservoir of an infectious agent and (b) transmission of a pathogen?

ANSWERS

Completion

1. less; 2. pandemic; 3. carriers; 4. morbidity; 5. animals; 6. milk pasteurization; 7. AIDS-related-complex; 8. propagated; 9. herd immunity; 10. carriers; 11. antibiotic resistant; 12. epidemiology; 13. 1855, contaminated water; 14. nosocomial; 15. fomite.

Multiple choice

1. B; 2. E; 3. D; 4. B; 5. D; 6. E; 7. A; 8. B; 9. E; 10. A. 11. C; 12. E; 13. A; 14. A; 15. C.

Discussion

1. See text Section 22.1

2. See text Section 22.3

3. See text Section 22.3

4. See text Section 22.4

5. See text Section 22.5

6. See text Section 22.5

7. See text Section 22.5

8. See text Section 22.6

9. See text Section 22.7

10. See text Section 22.8

MAJOR MICROBIAL DISEASES

OVERVIEW

Chapter 23 (pages 933-990) concerns the subset of the microorganisms that are pathogens. Infectious diseases are caused by bacteria, viruses, fungi, and protozoa. However, all pathogens which infect the same body tissue must overcome similar problems in initiating a disease in that tissue. Therefore, the discussion of pathogens in this chapter is organized with an emphasis on the disease's **mode of transmission** and mode of action.

CHAPTER NOTES

Microbes do not grow in **air**, but are transmitted through it, especially on dust particles. However, the organism must be relatively resistant to drying to be effectively transmitted. In general, Gram-positive bacteria such as *Staphylococcus* and *Streptococcus* survive longer under dry conditions than do Gram-negative species. Respiratory pathogens are released in **aerosols** by an infected host by sneezing, coughing, or talking. The density of pathogens in air will be greater in crowded indoor areas than outdoors.

Most bacterial pathogens of the respiratory tract are Gram positive species. They are susceptible to antibiotics such as penicillin, and effective vaccines can be constructed against them.

Streptococcus pyogenes is frequently part of the normal flora in the nasopharynx of healthy adults. However, low host resistance may result in strep-throat, with inflammation of the throat and a mild fever. Accurate diagnosis is important, because the bacterial infections can be treated with penicillin.

Some *S. pyogenes* strains carry a lysogenic bacteriophage which codes for the exotoxin that produces the rash of **scarlet fever**. If streptococcal infections are not completely eradicated, an **autoimmune** condition such as **rheumatic fever** may arise. Antibodies made against the streptococcal cell surface antigens may cross-react with heart and joint tissue, or cause kidney damage.

Streptococcus pneumoniae, a cause of bacterial **pneumonia**, is a member of the normal upper respiratory tract flora in many individuals. However, if it infects the lungs, a strong inflammatory response occurs which impairs lung function. Antiphagocytic capsules are an important virulence factor of the pathogen. The infection responds to penicillin; if untreated the infection may be fatal. *Staphylococcus aureus* is also commonly found in humans, particularly in the upper respiratory tract. However, strains of *S. aureus* have been linked to toxic shock syndrome (TSS).

Diphtheria is caused by *Corynebacterium diphtheriae*. The organism infects the throat and tonsils; **neuraminidase** may enhance its invasiveness. Its growth forms a **pseudomembrane** which makes breathing difficult. However, systemic effects are the result of an exotoxin which inhibits eukaryotic protein synthesis. The genetic information for the toxin is located on a prophage. Infection is prevented by immunization with DPT vaccine. Infected individuals are treated with antibiotics and antitoxin.

Legionella pneumophila can cause pneumonia in **compromised** hosts, such as the elderly. The organism is a normal inhabitant of many natural environments. It is transmitted to humans in **aerosols** generated from air conditioning units, where it can grow in cooling water.

Infants are susceptible to **whooping cough**, caused by *Bordetella pertussis*. Bacteria adhere to cells of the upper respiratory tract using a surface component called **filamentous**

hemagglutinin antigen. The bacterium produces an exotoxin which promotes synthesis of cyclic AMP. The symptoms are a violent cough. A vaccine of killed cells is available.

Mycobacterium tuberculosis is responsible for the lung disease **tuberculosis**. Primary infection occurs by inhalation of droplets containing the bacteria. The microbes grow in the lungs, and induce a cell-mediated immune response which involves a **delayed hypersensitivity** reaction. The aggregation of activated macrophages around the focus of infection forms a **tubercle** which walls off the infection. However, bacteria remain viable in tubercles for years. If the host is stressed, these bacteria may be released and the lung infection will spread. To identify people who have been exposed to tuberculosis, **tuberculin** protein is injected under the skin; this produces a localized delayed-hypersensitivity reaction in individuals whose immune systems have responded to the presence of *M. tuberculosis*. Tuberculosis can be treated with the drug **isoniazid**, which blocks synthesis of a lipid unique to mycobacteria. Treatment must be continued for long periods to eliminate all pathogens embedded in tubercles.

Tuberculosis can also be caused by another mycobacterium, *M. bovis*, which is transmitted through milk from infected cattle. **Pasteurization** of milk has eliminated this threat. Another pathogenic mycobacterium is *M. leprae*. It is not a respiratory pathogen, but grows intracellularly in macrophages in the skin and causes **leprosy**. The disease is not highly contagious, and is transmitted by direct contact. Infections are more common in the tropics.

There are several viral infections of the respiratory tract. The most prevalent is the common cold, caused by **rhinoviruses**. No effective long-term immunity can be generated because there are more than 100 serologically different rhinoviruses. Because there are few antiviral drugs available, little can be done to treat the infection.

Influenza is transmitted from human-to-human through air. It is caused by influenza virus, which infects the mucous membranes of the upper respiratory tract. Symptoms of fever, chills, fatigue and headache last 3 to 7 days, and then recovery is rapid. In individuals with weakened defense mechanisms, (infants and the elderly), secondary infections such as bacterial pneumonia may occur, and these infections may be fatal. Influenza occurs in **pandemics** when a new virus strain of high virulence and novel antigenic structure arises. **Antigenic shifts** in the external protein coat arise because the **segmented** virus genome can reassort in cells infected with more than one virus. Modifications in **hemagglutinin** and **neuraminidase** proteins seem especially important in this regard, because immunity to influenza depends upon IgA antibodies against these proteins.

Several childhood diseases of viral origin are transmitted through air. Measles is caused by **rubeola** virus. A combination of humoral and cellular immunity eliminates the virus in about ten days. Mumps, like measles, is caused by a **paramyxovirus**. Rubella is due to a **togavirus**. The virus is transmitted from mother to fetus during the first trimester of a pregnancy, with severe effects. **MMR vaccine**, containing live attenuated viruses, provides protection against these three diseases. **Varicella** virus, of the Herpes group, causes chickenpox. The virus is disseminated by the bloodstream and causes a systemic rash. After recovery, virus may exist in nerve cells in **latent** form. A subsequent active infection of skin cells causes **shingles**.

Respiratory infections are difficult to control because there is no practical way to prevent transmission through air. Furthermore, some of the pathogens reside in healthy carriers, and therefore are widely disseminated.

Sexually-transmitted diseases (STD) are caused by a cross section of bacteria, viruses, and protozoa. Public health control is difficult because it requires information on sexual practices that are sensitive issues in society. All these diseases are transmitted by intimate direct contact. Note that many of them can be transmitted to newborns from infected mothers during birth.

Gonorrhea now occurs in epidemic proportions. The bacterium *Neisseria gonorrhoeae* infects mucous membranes, usually in the genitourinary tract. Infections in females often go unnoticed; therefore they can serve as carriers to disseminate the disease. In males, the infection causes a painful inflammation of the urethra. In the past, penicillin was effective in curing gonorrhea. However, penicillin-resistant strains harboring a plasmid which specifies a penicillinase are becoming prevalent. Infection does not confer protective immunity; therefore, reinfection is possible.

Syphilis is caused by the obligate anaerobe spirochete *Treponema pallidum*. There are three stages to the disease. After infection through breaks in the epidermis, the **primary lesion** is a **chancre** at the site of entry, usually on the genitalia. This disappears, but some microbes disseminate through the body to mucous membranes, joints, and the central nervous system. The **secondary stage** is a skin rash, caused by a hypersensitive reaction to the treponeme. In both stages, the individual is infectious. In the absence of penicillin treatment, some individuals enter the **tertiary stage**, in which severe symptoms in the cardiovascular or central nervous systems can occur due to hypersensitivity reactions.

The obligate intracellular parasite *Chlamydia trachomatis* is responsible for many cases of **nongonococcal urethritis**. The occurrence of this disease may exceed that of either gonorrhea or syphilis. This results because *C. trachomatis* causes an inapparent infection, so that unknowing carriers may disseminate the organism. Immunological tests are now used to diagnoses the occurrence of the organism. *C. trachomatis* is also the causative agent for lymphogranuloma venereum a disease which occurs most frequently in males.

Herpesviruses cause latent infections in humans. Those of type 1 cause cold sores, whereas type 2 cause blisters in the genital region. Genital herpes is incurable, but the blisters can be controlled with **acyclovir**.

The protozoan *Trichomonas vaginalis* can cause nongonococcal urethritis. In this instance, males are important asymptomatic carriers transmitting the disease. Transmission can be prevented by the use of condoms; metronidazole is used to treat infections.

HIV virus is the agent of the **fatal disease AIDS**. Fatalities result because the virus impairs the host's immune system, so that the host is susceptible to **opportunistic infections** by microbes that are rarely invasive in healthy humans. These include the protozoan *Pneumocystis carinii*, toxoplasmosis, and systemic yeast infections.

HIV virus specifically infects the CD4 class of **T-lymphocytes**, cells involved in the immune response. As a result, the end result of an HIV infection is a loss of CD4 cells. As the CD4 makes up about 70% of the T-cells, their loss causes the susceptibility to the opportunistic infections mentioned above. T cell function is lost because (a) cells producing virus stop dividing and (b) uninfected T cells become nonfunctional when they bind to infected cells and fuse to form **syncytia**. Overall lymphocyte function declines as the level of **immune modulators** produced by CD4 cells decreases.

While no cure is available for HIV infections, several drugs are reported to delay the onset of AIDS. Azidothymidine (AZT) is most commonly used drug and inhibits HIV replication. The genetic variability in the HIV virus has slowed the development of an effective vaccine.

Rabies is a **zoonosis** that can be transmitted to humans by an animal bite. It is caused by a virus of the **rhabdovirus** family, which attacks the central nervous system and if untreated, causes death. The incubation period is about two weeks in dogs, but up to nine months in humans. The virus proliferates in the brain; death is due to respiratory paralysis. Vaccines of attenuated live virus are available; because of the long incubation period, they can be administered after infection. An infected individual will also be **passively immunized** with antiserum. The reservoir of rabies virus is the wild animal population.

Rickettsial infections (**typhus fever** and **Rocky Mountain Spotted fever**) are transmitted by insect bites. The initial symptoms are fever, headache, and a rash. These bacteria are obligate intracellular parasites of phagocytic cells. Because rickettsias are not easily cultured, diagnosis employs immunological tests. These diseases are controlled by control of the insect vectors: lice, fleas, and ticks.

Lyme disease is transmitted to humans by tick bites. The disease is predominate in the upper Midwest but is spreading to all parts of the United States. It is caused by a spirochete, *Borrelia burgdorferi* which is transmitted to humans as the tick is feeding on blood. Although easily treated soon after infection, if the disease progresses to the **chronic** stage, severe **neurological** symptoms occur which are similar to the effects produced by the spirochete which causes syphilis.

Malaria is caused by protozoa of the Sporozoa group; *Plasmodium vivax* is the most widespread. A complete life cycle of this organism requires growth in both humans and mosquitoes of the genus *Anopheles*. Therefore, the distribution of the disease parallels the geographic distribution of *Anopheles*. The characteristic cycles of fever and chills are due to cycles of infection of red blood cells by **merozoites** and their release from these cells. Humans are infected by **sporozoites** that are formed in the mosquito; the mosquito is infected by **gametocytes** that arise in the human. Two drugs, chloroquine and primaquine, together eliminate parasites both inside and outside red blood cells. Disease control involves eliminating mosquitoes by either draining swamps which they inhabit or killing them with insecticides.

Humans who reside where malaria is endemic have evolved resistance mechanisms to *Plasmodium*. This may involve an altered **hemoglobin** with reduced affinity for oxygen in red blood cells. The parasites do not grow as well in these cells.

Yersinia pestis causes plague. Its reservoir is rodents. The bacterium is transmitted to humans via flea bites. The microbe infects the lymph nodes, which are then called **buboes**. A capsule and other surface components are important virulence factors which prevent phagocytosis. An exotoxin that inhibits respiration in mitochondria may be important in causing damage to the host. In this form of the disease, called **bubonic plague**, the bacteria cause a septicemia, and eventually death. If *Y. pestis* is inhaled, **pneumonic plague** results. This respiratory infection is highly contagious and fatal if untreated. Rapid diagnosis and antibiotic therapy can prevent death by this pathogen.

Foodborne disease can either be **food poisonings**, which result from ingestion of bacterial exotoxins that were synthesized in the food, or **food infections**, caused by growth in the body of bacteria that contaminated the food.

Staphylococcus aureus can grow in creamed foods that are not properly refrigerated, and produce a heat-stable **enterotoxin**. Vomiting and diarrhea occur a few hours after it is ingested.

Clostridium perfringens proliferates in meats left in warming trays. The bacteria which are ingested sporulate in the intestine, and during this process release an enterotoxin which induces diarrhea 8-22 hours after the contaminated food was consumed.

Clostridium botulinum is responsible for a fatal food poisoning. A heat-sensitive exotoxin is produced when the microbe grows in food which has been inadequately processed to kill endospores. If the food is not subsequently cooked, the toxin will cause paralysis.

Salmonella species may contaminate meat, poultry, and dairy products. The source of contamination is either the animal itself or food handlers. If the food is not adequately cooked before eating, the viable bacteria can infect the intestinal tract, and induce fever, vomiting, and diarrhea.

Campylobacter species cause bacterial diarrhea in children. Poultry is a major reservoir of the pathogen. It can also be transmitted by contact between children and dogs.

Infectious **hepatitis** is a viral disease transmitted via fecal contamination of water, food, or milk. The virus spreads from the intestine through the bloodstream to the liver, where it causes jaundice. Shellfish harvested from water contaminated with human feces are a major vehicle for the disease. Hepatitis types B and C also affect the liver. They are transmitted through blood.

Escherichia coli strains can cause diarrhea. This trait is plasmid-encoded (see text section 11.9). Disease occurs in travelers who have not encountered these strains previously; the local population has developed immunity due to IgA antibodies which prevent colonization of the intestinal mucosa.

Many intestinal pathogens leave the body in the feces. If **drinking water** supplies become contaminated with these feces, the pathogens can be transmitted to susceptible hosts. Widespread consumption of contaminated drinking water will lead to disease **epidemics**. In developed countries, the transmission of waterborne pathogens has been controlled by treatment of sewage, and purification of drinking water.

The most serious waterborne bacterial disease is **cholera**, caused by *Vibrio cholerae*. Cholera enterotoxin induces a severe diarrhea, which can result in loss of 20 liters of fluid per day and death by dehydration. However, fluid replacement permits survival of the host. Cholera still occurs where sewage is not properly treated. A large inoculum of bacteria must be ingested so that some cells survive the acidity of the stomach. These microbes attach to the epithelium in the small intestine and produce enterotoxin.

The protozoan *Giardia lamblia* causes gastrointestinal disease. A resting stage called a **cyst** is transmitted through water contaminated with feces of wild animals. The cysts germinate in the intestinal tract to form **trophozoites** which cause diarrhea. Unlike most waterborne pathogens, *Giardia* is resistant to chlorine treatment. The cysts can be removed by sedimentation or filtration of drinking water.

Entamoeba histolytica is an **anaerobic** protozoan which is transmitted as cysts in water. The growth of trophozoites that arise from cysts on and in intestinal mucosal cells causes **ulceration** of the mucosa and **dysentery**. If untreated, the protozoa may infect other body organs.

Water treatment can prevent the spread of waterborne pathogens in water supplies. However, it is often technically difficult to directly detect these pathogens in water. Therefore, **indicator** bacteria are used to detect fecal contamination of water. The **coliform** group is used, because they inhabit the intestinal tract of humans and warm-blooded animals, are present there in large numbers, and persist in natural and treated water in a similar way as actual intestinal pathogens. If high numbers of coliforms are found in a water supply, it indicates that the water has been contaminated with feces, and that it potentially (but not necessarily) contains intestinal pathogens. Note that the term "coliform" is an operational definition, and includes several species, not all of which are strictly intestinal bacteria.

The number of viable coliforms is quantified in water using a selective medium to exclude the growth of other organisms. To be considered safe for drinking, water must contain on average less than one coliform per 100 ml.

Drinking water is treated to remove pathogens, turbidity, color, and odor. Treatments differ with individual circumstances, but in general they include **sedimentation**, **filtration**, and **chlorination**. Sedimentation of particles may be accelerated by adding a **coagulant** to form large precipitates which settle more quickly. Sand filters remove most particles and microbes which did not sediment. To assure microbiological safety, **chlorine** is added to the water supply to kill microorganisms.

While bacteria, viruses and protozoa are important pathogens, fungi can cause a number of diseases. Fungi are primarily opportunistic pathogens, tending to attack medically compromised individuals. Fungi cause disease in three ways: production of an allergic or **hypersensitivity** reaction, production of toxic exotoxin termed **mycotoxins** and actual invasion and growth in the body termed **mycosis**.

SELF TESTS

TRUE OR FALSE

1. Throat and skin infections due to *Staphylococcus aureus* are prone to cause rheumatic fever.

2. The toxin of *Corynebacterium diphtheriae* acts at a very specific site in protein synthesis.

3. Gastroenteritis due to *Salmonella* strains is commonly food-borne.

4. Botulism can be caused by the ingestion of improperly processed canned food.

5. Circulating antibody is of primary importance in immunity to rhinoviruses.

6. Diseases of animals are not transmitted to man.

7. The causative agent of syphilis is *Treponema pallidum*.

8. Deaths following influenza are usually due to pneumonia.

9. The cause of typhoid fever is *Salmonella typhi*.

10. Botulism results from the infection of the large intestine by *Clostridium botulinum*.

11. Different microbial species may produce the same disease symptoms.

12. The same microbial species may cause more than one disease.

13. The carrier state is rare following infection with *Neisseria gonorrhoeae*.

14. Sexually-transmitted diseases are currently under control because of strict public health surveillance.

MATCHING

Match the disease with its mode of transmission.

Disease	Mode of Transmission
1. tuberculosis	(A) direct contact
2. whooping cough	(B) air
3. cholera	(C) water or food
4. infectious hepatitis	(D) insects
5. rabies	(E) animals
6. gastroenteritis	
7. herpesvirus type 1 infection	
8. herpesvirus type 2 infection	
9. Rocky Mountain spotted fever	
10. gonorrhea	
11. mumps	
12. malaria	
13. amoebic dysentery	
14. legionellosis	
15. bubonic plague	

COMPLETION

1. The antibiotic of choice in streptococcal infections is _____.

2. The _____ of *Streptococcus pneumoniae* enables it to evade phagocytosis.

3. *Legionella* species cause disease by infecting the _____.

4. Measles, mumps, and rubella can be prevented by _____.

5. New, virulent forms of influenza virus arise periodically because of antigenic shifts in the _____ of the virus.

6. Infections of _____ are usually asymptomatic in males.

7. The drug acyclovir is useful in treating the blister stage of _____ infections.

8. Death of AIDS patients is generally due to _____.

9. HIV virus specifically infects _____.

10. _____ virus proliferates in the central nervous system.

11. _____ are bacterial pathogens which are obligate intracellular parasites, and which are transmitted between hosts by insects.

12. Resistance to malarial infection is enhanced by an altered form of_____.

13. The reservoir of the plague bacterium is _____.

14. The symptoms of perfringens food poisoning are caused by _____.

15. _____ is a major cause of bacterial diarrhea in children.

16. _____ is a protozoal parasite which is transmitted through water.

17. Fecal contamination of water supplies is indicated by the presence of_____ bacteria.

KEY WORDS AND PHRASES

β-hemolysis (p 936)	airborne transmission (p 934)
antigenic shift (p 947)	azidothymidine-AZT (p 951)
chancre (p 954)	chlamydia (p 954)
chlorination (p 982)	congenital syphilis (p 954)
cyst (p 985)	DPT vaccine (p 939)
dysentery (p 982)	enterotoxin (p 979)
gametocyte (p 972)	histoplasmosis (p 957)
HIV virus (p 957)	impetigo (p 938)
Kaposi's sarcoma (p 951)	legionellosis (p 939)
membrane filter procedure (p 986)	merozoite (p 971)
MMR vaccine (p 949)	most probable number (p 986)
mycoses (p 987)	neuraminidase (p 939)

opportunistic infection (p 988)	pandemic (p 947)
pasteurization (p 944)	pertussis (p 941)
pneumonia (p 936)	pseudomembrane (p 939)
Q fever (p 967)	reverse transcriptase (p 958)
rheumatic fever (p 936)	rickettsia (p 966)
salmonellosis (p 979)	scarlet fever (p 936)
Schick test (p 939)	sporozoite (p 971)
toxic shock syndrome (p 938)	trophozoite (p 985)
typhoid fever (p 979)	

MULTIPLE CHOICE

1. The elimination of epidemics of typhoid fever in the United States has been due to:

   (A) control of mosquito populations

   (B) treatment of sewage and water supplies

   (C) a decrease in virulence of the causative agent

   (D) antibiotic therapy

   (E) vaccination

2. The disease shingles occurs in individuals

   (A) infected with *Bordetella pertussis*

   (B) infected by HIV virus

   (C) in which varicella virus persisted after a childhood case of chicken pox

   (D) that have a deficient immune system

   (E) that have not completely overcome strep throat

3. Complications of strep throat can be prevented by:

   (A) prompt diagnosis and adequate treatment of b-hemolytic streptococcal infections

   (B) frequent antiseptic mouth washes

   (C) a high calorie diet

   (D) bed rest

   (E) none of the above

4. An effective vaccine against the common cold does not appear feasible because:

   (A) there are over 100 different rhinoviruses

   (B) there is no immunity to rhinoviruses

   (C) rhinoviruses cannot be cultivated in culture

   (D) rhinoviruses cannot be attenuated for safe use

5. The importance of rubella virus is that it:

    (A) is a frequent cause of viral enteritis

    (B) can cross the placental wall and infect the fetus

    (C) can cause birth defects if fetal infection occurs early in a pregnancy

    (D) B and C above

    (E) all of the above

6. Which of the following represents the greatest potential source of pathogens in municipal sewage?

    (A) industrial wastes

    (B) domestic (home) wastes

    (C) storm sewers

    (D) rainfall runoff

7. The secondary stage of syphilis is characterized by

    (A) insanity

    (B) arthritis

    (C) chancre

    (D) skin rash

    (E) diarrhea

8. Which of the following diseases result in no immunity after recovery?

    (A) chicken pox

    (B) gonorrhea

    (C) cholera

    (D) A and B above

    (E) all of the above

9. Exotoxins are responsible for the symptoms of all of the following diseases EXCEPT:

    (A) diphtheria

    (B) botulism

    (C) tuberculosis

    (D) *Staphylococcus aureus* food poisoning

    (E) cholera

10. Why are there so many more female carriers of gonorrhea than male carriers?

    (A) The disease has no serious consequences in the female.

    (B) The symptoms in the female are much less noticeable.

    (C) The organism will not grow at the pH of the vagina.

    (D) The organism does not produce a chronic infection in the male.

    (E) Females are more promiscuous.

11. An individual who experiences vomiting, nausea, and abdominal cramps four hours after eating potato salad probably has:

   (A) salmonellosis

   (B) botulism

   (C) staphylococcal food poisoning

   (D) perfringens food poisoning

   (E) trichinosis

12. The most important method of **preventing** diphtheria is:

   (A) avoiding infected persons

   (B) immunization with diphtheria toxoid

   (C) immunization with diphtheria antitoxin

   (D) prescription of antibiotics

13. Rheumatic fever results from:

   (A) invasion of bones and joints by *Mycobacterium tuberculosis*

   (B) invasion of heart and joints by an exotoxin of *Corynebacterium diphtheriae*

   (C) a hypersensitivity reaction to an antigen of *Streptococcus pyogenes*

   (D) heart valve hypersensitivity to influenza virus

   (E) a delayed reaction to *Micrococcus rheuminantum*

14. Influenza

   (A) is a nuisance, but never leads to death in healthy people

   (B) can cause death in certain groups, such as the elderly

   (C) has been controlled by universal immunization with live vaccine

   (D) is caused by several different bacteria and viruses

   (E) viruses have historically been genetically stable

15. Q fever is caused by an intercellular parasite, *Coxillea burnetii* which is often associated with cattle. How is Q fever spread:

   (A) by cattle ticks transmitting the disease in herds

   (B) by improperly pasteurized milk transmitting the disease to humans

   (C) by rainfall runoff washing the bacteria into water supplies

   (D) by domestic (home) wastes leaching the bacteria into water supplies

   (E) two of the above

DISCUSSION

1. What characteristics must a pathogen possess to be transmitted through (a) air, (b) water, or (c) animals? Which pathogens are only transmitted by direct contact and why are they restricted to this mode?

2. Why can pathogenic streptococci be part of the normal flora in many adults, yet cause disease in others?

3. Contrast the symptoms, treatment, and long-term effects of lung infections caused by *Streptococcus pneumoniae*, *Legionella pneumophila* and *Mycobacterium tuberculosis*.

4. Which types of respiratory infections are most common? Which are most difficult to treat with drugs? Which are most likely to cause death if untreated?

5. What public health measures can be employed to control the incidence of (a) respiratory infections, (b) sexually-transmitted diseases, and (c) water-borne diseases?

6. What elements of the immune system are rendered nonfunctional by HIV virus?

7. What characteristics of the various sexually-transmitted diseases make these infections difficult to control in the human population?

8. What techniques are used to (a) diagnose and (b) control infections caused by rickettsias?

9. *Yersinia pestis* was responsible for killing one-fourth of the population of Europe in the 14th Century (the "black death"). What environmental conditions could have promoted this pandemic?

10. What procedures should be undertaken in the selection, preparation, and storage of food to minimize foodborne diseases?

ANSWERS

True and false

1. False; 2. True; 3. True; 4. True; 5. False; 6. False; 7. True; 8. True; 9. True; 10. False; 11. True; 12. True; 13. False; 14. False.

Matching

1. B; 2. B; 3. C; 4. C; 5. E; 6. C; 7. B; 8. A; 9. D; 10. A; 11. B; 12. D; 13. C; 14. B; 15. D.

Completion

1. penicillin; 2. capsule; 3. lungs; 4. vaccination; 5. protein coat; 6. *Trichomonas vaginalis*; 7. genital herpes; 8. opportunistic infections; 9. T4 lymphocytes; 10. rabies; 11. rickettsias; 12. hemoglobin; 13. rodents; 14. exotoxin; 15. *Campylobacter*; 16. *Giardia lamblia*; 17. coliform.

Multiple choice

1. B; 2. C; 3. A; 4. A; 5. D; 6. B; 7. D; 8. B; 9. C; 10. B; 11. C; 12. B; 13. C; 14. B; 15. E.

Discussion

1. See text Sections 23.1, 23.6, 23.11, and 23.14

2. See text Section 23.2

3. See text Sections 23.2 and 23.3

4. See text Sections 23.2, 23.3, 23.4 and 23.5

5. See text Sections 23.5, 23.6, and 23.14

6. See text Section 23.7

7. See text Section 23.7

8. See text Section 23.9

9. See text Section 23.11

10. See text Section 23.13